# Christian Churches and Nigeria's Political Economy of Oil and Conflict

# Christian Churches and Nigeria's Political Economy of Oil and Conflict: Baptist and Pentecostal Perspectives

By

Nkem Emerald Osuigwe

CAMBRIDGE SCHOLARS

PUBLISHING

Christian Churches and Nigeria's Political Economy of Oil and Conflict:
Baptist and Pentecostal Perspectives,
by Nkem Emerald Osuigwe

This book first published 2014

Cambridge Scholars Publishing

12 Back Chapman Street, Newcastle upon Tyne, NE6 2XX, UK

British Library Cataloguing in Publication Data
A catalogue record for this book is available from the British Library



ISBN (10): 1-4438-6123-5, ISBN (13): 978-1-4438-6123-6

To the memory of my late parents and all the *innocent victims*
of the Niger Delta conflict

# TABLE OF CONTENTS

# LIST OF FIGURES AND TABLES

# FOREWORD

It is my pleasure to commend this work by Nkem Osuigwe as a fascinating insider account of the effects of oil related corruption, pollution and violence on the communities and churches of the Niger Delta. The corruption and neglect that have characterised the actions of the oil industry in the Niger Delta over more than fifty years are beginning to be exposed in the Nigerian and Western media but this study adds a valuable ethnographic depth to the impacts of the industry in the region. As is increasingly well known, the Delta has endured a vast number of leaks from poorly maintained overground pipes left unrepaired even as they pass through crop lands, forests and even villages, and numerous unchecked leaks from well heads. The large ecosystem of the Delta as well as human dwellings have been systematically polluted and no proper clean-up has been attempted, despite numerous attempts to force the oil companies to clean up through civil protests and in the courts. To add insult to injury, as the peoples of the Delta have seen their lands, waters and air polluted and wasted by the oil companies, they have not benefited much from the oil revenues that have gone into government coffers and oil company profits. Much is made in the world media of the desolation by the oil industry of the waters around North America and other developed countries from major oil leaks such as the Deepwater oil blowout in the Gulf of Mexico in 2010. But the oil spilled in the Niger Delta far exceeds that massive blowout both in quantity and in the effects on people and the environment. It has however not attracted the same kind of attention and the companies have neglected their responsibilities to stop leaks and clean up the oil in affected areas. As is so often the case, environmental injustice and pollution, coupled with natural resource exploitation by powerful outside interests, have together occasioned growing civil conflict.

The Niger Delta has been plagued by oil-related violence for years. Civil protests have been violently put down by security personnel and police and the prominent literary voice of the oil protest in the Delta of Ken Saro Wiwa was brutally murdered in jail. The violent response of the authorities to legitimate civil protest has in turn provoked an alternative response in the region which has also become a major cause of violence. As the people of the Delta have been denied benefits from the oil while their environment has been ruined by the oil industry, some have taken affairs into their own hands. There is now widespread pilfering of oil from wellheads and pipelines which has magnified the problem of oil spillages. At the same time there has been violent conflict between gangs involved in extracting and marketing stolen oil obtained which has seen a tragic rise in gun-related injuries and killings particularly amongst young men in the region.

The Christian Churches have a more than one hundred year history of mission and witness in the Niger Delta and since many in the region attend church, many church members have been affected by the corruption and pollution of the oil industry in the region, and the civil conflict that it has provoked. In this study Osuigwe has used his position as a Baptist pastor in the region, and his knowledge of his own and other churches, to unfold an ethnographic study of the effects of and responses to the oil-related violence among congregations and their members. The study gives a powerful insight into the ways in which Christians in the region have responded to the violence, through forms of worship, through attempts at reconciliation and encouragement to the participants to lay down their arms, and through attempts at interventions with the authorities to address the roots of the violence in the structural injustices of the oil industry in the Delta. This book also makes an important contribution to the scholarly study of African independent churches. The received view amongst Western scholars is that African evangelical churches are conservative and quietist in their relationships to the wider society and that they largely reject the attempts of the more established mission churches to link Christian mission and worship with African cultural concepts and practices. This study shows that on the contrary such churches in the Delta have been much involved in societal efforts to address and repair oil-related violence and injustice, and that the faith and worship of their members and pastors contextualise the gospel in a range of ways in the cultural idioms and practices of the region.

I commend this book warmly to its potential readers. Osuigwe has a powerful story to tell which includes much human suffering from which his own church and family has not been immune. He tells it engagingly and passionately while at the same time showing the scholarly implications of his story for the study of African Christianity and for African Christian theology as well as for the political economy of oil in the developing world

**Professor Michael Northcott**
Professor of Christian Ethics, School of Divinity
University of Edinburgh

# LIST OF ABBREVIATIONS

| | |
|---|---|
| AICs | African Instituted (or Independent, Initiated, Indigenous) Churches |
| bpd | barrels per day |
| BSF | Baptist Students Fellowship |
| CAN | Christian Association of Nigeria |
| CCN | Christian Council of Nigeria |
| CfAN | Christ for All Nations |
| CMS | Church Missionary Society |
| C of O | Certificate of Occupancy |
| COMAND | Coalition of Militant Action in the Niger Delta |
| CPFN | Christian Pentecostal Fellowship of Nigeria |
| ECWA | Evangelical Church of West Africa [now Evangelical Church Winning All] |
| EDICESA | Ecumenical Documentation and Information Centre of Eastern and Southern Africa |
| ERA | Environmental Rights Action |
| FoEN | Friends of the Earth Nigeria |
| GRA | Government Reservation Area |
| HRW | Human Rights Watch |
| H2H | Heart-to-Heart |
| INC | Ijaw National Council |
| IVF | Idumini Volunteer Force |
| IYC | Ijaw Youth Congress |
| JRC | Joint Revolutionary Council |
| JTF | Joint Task Force |
| MB | Martyrs Brigade |
| MEND | Movement for the Emancipation of the Niger Delta |
| MNOCs | Multi-national Oil Companies |
| MOPOL | Mobile Police |
| MOSOP | Movement for the Survival of Ogoni People |
| MoU | Memorandum of Understanding |
| NDDB | Niger Delta Development Board |
| NDDC | Niger Delta Development Commission |
| NDPKVF | Niger Delta Peace Keeping Volunteer Force |
| NDPVF | Niger Delta Peoples Volunteer Force |

| | |
|---|---|
| NDVS | Niger Delta Vigilante Service and Niger Delta Vengeance Seekers |
| NGO | Non-Governmental Organisation |
| NEITI | Nigeria Extractive Industries Transparency Initiative |
| NNOC | Nigerian National Oil Corporation |
| NNPC | Nigerian National Petroleum Corporation |
| NT | New Testament |
| NUPENG | National Union of Petroleum and Natural Gas Workers |
| OAIC | Organisation of African Instituted Churches |
| OMPADEC | Oil Mineral Producing Areas Development Commission |
| ORA | Operation Reach All |
| PA | Personal Assistant |
| PENGASSAN | Petroleum and Natural Gas Senior Staff Association of Nigeria |
| PFN | Pentecostal Fellowship of Nigeria |
| SOAS | School of Oriental and African Studies (University of London) |
| SPDC | Shell Petroleum Development Company |
| TEKAN | *Tarayar Ekklesiyoyin Kristi a* Nigeria |
| WAEC SSCE | West African Examination Council Senior Secondary Certificate Examination |

# ACKNOWLEDGEMENTS

This book grows out of my 2010 University of Edinburgh PhD thesis entitled "Crude Oil, Conflict and Christian Witness in Nigeria: Baptist and Pentecostal Perspectives." The support and assistance of several people towards the completion of the study is worthy of note. I have categorised such support and assistance into six overlapping areas, based on the clusters of challenges I faced during the course of my programme. The areas include *academic*, *pastoral*, *spiritual*, *moral*, *financial* and *practical*. Some individuals and groups fall into two or more clusters, while others fall into one. Yet, there may be others who could have played behind-the-scene roles, choosing to be anonymous, and whom I might never know. But whether there are clusters, and whether these are anonymous or not, my deepest gratitude goes to each one of them.

First, I must appreciate my church, Faith Baptist Church, Port Harcourt and the entire membership for their spiritual, moral and financial support. Without their gracious approval of study leave for me, I would not have been able to embark on this journey in the first place. Worthy of note in this regard are the Pastoral Team, the Church Council, and the Diaconate for leading the church in my absence and for mobilising the members to give financially towards my tuition fees. This was especially so in 2008/2009 when, as a result of the paralyses of economic activities in Port Harcourt occasioned by youth militancy, I had to suspend my studies for one year. Thus the support – spiritual, moral and financial – which I received from the members is highly appreciated. I must not fail to thank especially the mothers in the church under the auspices of the Women's Missionary Society (WMS) for their collective financial support in 2007 and 2009. God bless you, mothers! I am grateful.

In addition, there are individual members of the church who supported me financially. They are too numerous to mention, but just a few: Deacon and Deaconess Sunday Olutayo-Irheren, Deacon and Deaconess Justice Chuku, Dr Daru Owei, Mrs. Mina Bob-Manuel, Rev. and Rev. Mrs. Isaac Ibude, Mr and Mrs. Ariko, and Mr and Mrs. Charles Akah. Others include Deacon Prof. and Deaconess Nkem Okoh, Deacon and Mrs. Isobo Jim-George, Mr and Mrs. Isoboye Warmate, Eng. and Dr Mrs. John Oyeniyi, Rev. and Mrs. Robert Okparanta, Mr. Dennis Ogboru, and Deacon and Mrs. A.A.O. Abolarin. Not left out are Deacon and Deaconess G. Chuku,

Mr and Mrs. Mutiu Sunmonu, Mr and Mrs. Eze Ugoji, Mr and Mrs. Fubara Batubo and Mrs. Sarah Okoro.

Yet there are others who freely gave their moral support by way of text messages and phone calls. They include Raphael Frank, Ma I.D. Lawson, Grace Affi, Rose Affi, Azubuike Joel, Collins and Mary Ochomma and their children, Mr and Mrs. John Fagbile, Victor and Gift Okoroma, Catherine Adagilo, Roseline Okere, Margaret Wachuku, Comfort Beako, Melford Agala, Louis Amadi, Peter Soberekon, Justina Soberekon, Brantley Izu, Rev. and Mrs. I.D. Amachree, Rev. Azubuike Ohaleta and Rev. and Mrs. Evangelist Abeke.

Next, I must express my deepest gratitude to my thesis supervisor, Prof. Michael Northcott, for his patience, drive and diligent guidance throughout the study. In fact, he proved to be more than just an academic adviser, as he became a worthy *research mentor*, *intellectual gadfly*, *cerebral midwife*, *academic critic* and *personal friend* in his quest to bring out the best in me regarding the subject matter of the study. I only hope that I have been able to meet the minimum of his expectations. Without his invaluable assistance in the clusters of academic, pastoral and moral support, this project would have been stillborn. Thank you very much. My appreciation in this regard also goes to Dr Afe Adogame for his critical feedbacks, which proved helpful in the final outcome of the research.

I am especially indebted to the leadership and entire membership of the three local congregations used in this study for their permission, cooperation and support. They not only accepted to be part of the study but willingly and enthusiastically participated through the focus group sessions, interviews and private conversations, despite the dangers associated with the inter-gang warfare that characterised the period. I also appreciate all those who served as interpreters and note-takers at the focus group sessions. Hilltop International Christian Centre, Port Harcourt, even took a step further by giving a token of their financial support, which is deeply appreciated.

On the practical cluster, I would be an ingrate if I fail to thank the administrative, IT and library staff at New College and the University Main Library for their support and assistance in friendly and professional ways when I called on them. Similarly, all the staff at the International Office of the University deserve my mention. Thank you for your assistance on the two occasions I needed to extend my student visa.

Of course, I must not fail to appreciate the leadership and membership of the African and Caribbean Christian Fellowship (ACCF), Edinburgh for their moral and spiritual support. Of particular significance in this regard are Dr and Mrs. Francis Fatoye, Dr and Mrs. Raphael Mrode, Dr and Mrs.

Samson Arugu, Mr. and Dr Mrs. Kingsley Obike and Rev. Dr and Mrs. Israel Akanji.

My gratitude also goes to the immediate past General Secretary of the Nigerian Baptist Convention, Rev. Dr Ademola Ishola, who led the Convention to approve some financial support for me on two different occasions in 2008 and 2010. I am grateful. I must also thank Mr. and Mrs. Vincent Hejirika in Epping, Essex, London for their moral support. On a couple of occasions they hosted me in their home while on my way to and/or from Nigeria for field research. Their hospitality and phone calls are deeply appreciated.

To members of my extended family, I say a big *Thank you* for your support: my dad (now late), Dn. G.A. Osuigwe, my brothers Uche, Philip, Azu, Chidi, as well as all their wives; then Onyewuchi and Kelechi, as well as my sister, Ogechi Onukaogu, and her family. I also appreciate my mother, Mrs. Theresa Osuigwe, posthumously, as she passed on to glory just two months before I proceeded to the UK for my doctoral studies. My gratitude as well goes to my late mother-in-law, Mrs. Charlotte Oruama, and my brothers-in-law, Tonipre Berena and Godwin Berena, for their support and encouragement.

Penultimately, I cannot quantify the level of support I received from my Sweetheart and Wife, Mercy Atonye Osuigwe. Her role during the period superseded the six clusters mentioned above: she was supporter, encourager, confidante, friend, financial manager and helper. She gallantly and gracefully managed our home in my absence, had to endure the pain of separation occasioned by my absence and, more importantly, did it all with equanimity and dignity. Also she ensured that I received money regularly by way of electronic international funds transfer. Thank you, my wife! Pearl and Jasper, my beloved daughter and son, respectively, you were just eight and four, respectively, when Daddy started the programme. Now you are fifteen and eleven, respectively! What a change! Thank you for showing your understanding despite the ache of my absence, as demonstrated by Jasper's regular query, 'When is Daddy coming back?' All these crystallised in pushing me to finish the race. I am grateful.

Above all else, may all glory and praise be to God Almighty, the Three in One, who made a way where there seemed to be no way, and in whom 'I live, move and have my being'! *Otuto diri Chineke!* (Glory be to God!).

# CHAPTER ONE

# BACKGROUND AND INTRODUCTION

## 1.1 Synopsis of Oil and Conflict in Nigeria

Nigeria produces about 3% of the world's crude oil and is Africa's leading oil producer, as well as the eleventh largest producer in the world (Energy Information Administration 2005). It should be noted that Angola at a point during the conflict overtook Nigeria as Africa's leading producer of crude oil. Oil accounts for about 80% of government revenue and 95% of export earnings (Omiyi 2004). Up to the 1960s, agriculture was the mainstay of her economy before crude oil assumed the position beginning from the 1970s, following the first wave of oil boom in 1973-74. From a meagre 5,100 bpd in 1958, when the country exported her first crude oil, the volume of Nigeria's crude oil export reached 2.5 million bpd in 2005. Also, the country's crude oil stock increased from an estimated 5 billion barrels in 1969 to more than 35 billion barrels by 2004 (OPEC 2004, 16/17), with the Government planning to extend it to more than 40 billion barrels by 2010 (Itsueli 2005, 9). Thus, within the first fifty years of oil exportation, Nigeria amassed nearly half a billion dollars in revenue.

Most of the crude oil comes from just nine of the thirty six states that make up the Nigerian Federation. Using oil deposit as the basis of classification, the nine states are politically and collectively known as the Niger Delta, which includes two states from the South-East geo-political zone (Abia and Imo), one from the South-West (Ondo), and six from the South-South (Akwa Ibom, Bayelsa, Cross River, Delta, Edo, and Rivers). The proportion of oil production from these states is uneven with three of them – Bayelsa, Rivers and Delta – accounting for about 75% of the entire production in the country and about 50% of Government revenue (African Network for Environment and Economic Justice, ANNEJ, 2004, 3).

Within this scenario, it is reasonable to state that nine states of Nigeria produce the bulk of the revenue with which the Nigerian Government at the three tiers runs its affairs. Despite the huge revenue from oil, the region from where the revenue derives is economically deprived and environmentally degraded. For example, Michael L. Ross, in a report for the UK Department for International Development notes as follows: "Poverty in the south-south region is 58.2 percent, the highest rate in Nigeria; literacy rates, access to health services, and access to safe water are exceptionally low, and unemployment rates are exceptionally high" (cf. Ross 2003).

The case of the communities from where oil is directly drilled is even more precarious. This is best illustrated by the experience of Oloibiri, the community where oil was first drilled and exported from Nigeria in 1956 and 1958, respectively. Emmanuel Emmanuel captures the situation succinctly:

> As of today, 48 years after, Oloibiri is referred to as fossil town because there is nothing to show that the town opened the door to the international oil market for Nigeria. The only historical relics there is an old signboard marked 'Oloibiri Oil Well 1' with over-grown bush. Oloibiri has no roads, no hospital, no electricity and no water supply after 48 years of oil record (Emmanuel 2004, 1).

For decades, especially during the 1970s and 1980s, the people of the region protested against what they considered to be the unjust way the Nigerian Government and the Multi-national Oil Companies (MNOCs) treated them. The protests were generally uncoordinated, muted and ill-organised (Ikelegbe 2001, 438). Thus, the Government and MNOCs, for mutually beneficial reasons (Frynas 2000, 8), chose and could afford to ignore their agitations. But things took a different turn in the early 1990s as a new wave of demand for economic equity and environmental responsibility came from the Ogoni through their umbrella body, the Movement for the Survival of Ogoni People (MOSOP) and under the leadership of the late Ken Saro-Wiwa. Saro-Wiwa and eight of his compatriots – Baribo Bera, Saturday Dbee, Nordu Eawo, Daniel Gbokoo, Barinem Kiobel, John Kpuinem, Paul Levura and Felix Nwate – were later executed in 1995 by the then ruling military junta in Nigeria. That act brought the nascent struggle to the international limelight.

The Ogoni example served as an impetus for other Niger Delta groups. Thus, beginning from 1999, when the military relinquished power to a democratically-elected civilian administration and thereby restored certain basic human rights, such as freedom of expression, the agitation became more vociferous, better organised, more widespread, and violent. The major purveyors of the violent form of the agitation are the youths. 'Youths' here refers to young people between the ages of sixteen and thirty five in various local communities in the region who either are organised to serve as the 'Youth wing' of their community development unions and/or ethnic bodies, or organise themselves as counter-force to their leaders so as to champion one cause or the other. They took up arms against the Nigerian Government, the MNOCs, and even against themselves, and began to fight for control/access to the lucrative but illegal sale of crude oil known as oil bunkering, which refers to the illicit pilfering and sale of crude oil siphoned from the network of oil pipelines that criss-cross the Niger Delta. They later resorted to hostage-taking and the seizure or destruction of oil facilities. The human and non-human costs of the violence are inestimable. It is estimated that between 1999 when civilian administration was restored and 2006, about 100,000 people lost their lives as a result of the violence (*The Guardian* 2006). There is hardly anyone in the region that has not been affected directly or indirectly by the violence. Government's response to the violence is mixed but has often been to use force and thereby elongate the circle of violence and exacerbate the reign of destruction in the region.

Several studies have been carried out from various perspectives and attempts have been made to address the problem. The perspectives include the legal, economic, human rights, policy, political science, and environmental. They focus mainly on describing and/or analysing the problems associated with oil in Nigeria or on proffering possible solutions to the socio-political, economic and environmental issues emanating from it. However, no study hitherto has been carried out on the problem from the perspective of Christian social witness. This book is designed to address this lack.

## 1.2 Overview of Christian witness in the Niger Delta

The first attempt at Christian witness in Nigeria was carried out by the Roman Catholics, who came from Portugal in the 16th century, first to Benin in 1515 and to Warri in the 1570s (Isichei 1995, 62-63; Sanneh 1983, 36-52). These initial efforts were deemed "futile, feeble, and spasmodic" (Ayandele 1966, 3).

However, this initial failure was reversed in the 19th century, especially between 1841 and 1891, when the major mission churches successfully entered the country. The missionary bodies of the period include: the Church Missionary Society (CMS), which came mainly from Germany, the Wesleyan Missionary Society, which came from England, the Foreign Mission Committee of the United Presbyterian Church of Scotland, the Foreign Mission Board of the Southern Baptist Convention, which came from America, and the Catholic Society of African Missions of France that came from France (Ajayi 1965, xiii-xiv).

Christian missions reached the Niger Delta during the period. Thus, the city states in the region embraced the (Anglican) faith: Bonny in 1864, Brass in 1868, New Calabar in 1874, and Okrika in 1880 (Ayandele 1966, 71). The Presbyterians had reached Old Calabar in 1846; the Methodists got to Opobo in 1884, the same year the Holy Ghost Fathers (Roman Catholic) arrived at Onitsha in Eastern Nigeria; the Qua Iboe Church was planted in Cross River in 1887 (McKeown 1902, 61-67); while the Baptists started their missionary work in Buguma City in 1893. Several of these missionary endeavours were carried out at the request and/or support of some of the local kings and chiefs. For example, the introduction of Christianity to Okrika was made possible with the aid of three Okrika chiefs, namely, Atorudibo, Ogan, and George (Ogan 1988, 22-23). Also, the planting of the Christian mission in Bonny in 1864 followed the request in 1861 of King William Dappa Pepple to the Anglican Bishop of London, after his reinstatement to the throne following a period of deposition and deportation to Fernando Po (Alagoa and Fombo 1972, 19-27; Ayandele 1966, 71-115). Some of the rulers and leaders played active roles in their churches, such as King George Pepple I of Bonny, who served as a Sunday School Teacher in his local congregation, and King Eyo Honesty VI of Creek Town in Old Calabar, who chaired the Finance Committee of his church (Ayandele 1966, 78 and 98, respectively).

According to G.D. Numbere, this early Christian missionary effort in the region was "not only spiritual but also social" (2006). Thus, despite the negative criticisms of their work, some of which are valid (Ayandele 1966, 83/84), the Missions carried out social services that positively impacted on the lives of the people. Their successes in this regard could be gauged by some of the roles they played: They helped to reduce some of the languages to written form, established schools and hospitals, and brought an end to such dehumanising practices as domestic slavery, human sacrifice, infanticide, twin murders, punishment by ordeal, immolation, and cannibalism. Most importantly, the Christian Missions stood for "social ideals such as brotherhood, freedom, individual rights, justice and honour," and Christianity became "a social leveller" (Ayandele, 83/84).

The rate of growth and success of these missionary bodies varied, but they later spread to different parts of the region. Eventually, beginning from the late 1960s, the Pentecostal fervour reached the area. Three indigenous Pentecostal leaders that sprang up in the area during this period include G.D. Numbere, who established the Greater Evangelism World Crusade in Rivers State in 1972 (Numbere 2002, 16), Mba Adozie in the southeast, and the more widely known late Benson Idahosa, who started the Church of God Mission (Numbere 2006). They are prefigured by

Garrick Braide, who was originally an Anglican. Following a visionary experience he had in 1912, he adopted a prophetic and healing ministry. However, as a result of suspicion and hostility from the Church, he and his followers pulled out in 1916. Some of his followers later formed the Christ Army Church (Isichei 1995, 287).

Christian witness in the region at present is represented by various church groups within the Christian Association of Nigeria. There are five blocks of churches in the body: the Roman Catholics; the Christian Council of Nigeria (CCN), comprising mainly of the Anglicans, Methodists, Presbyterians, and Baptists; PFN-CPFN (Pentecostal Fellowship of Nigeria and Christian Pentecostal Fellowship of Nigeria); TEKAN-ECWA (Evangelical Church of West Africa); and OAIC (Organisation of African Instituted Churches).

Getting the records of the current spread of these churches is very difficult. However, using Rivers State as an example, and based on information received from various sources, the configuration of some of these denominations could be deduced from the following data: the Roman Catholics have one Diocese in the State, the Port Harcourt Diocese, which is under the Calabar Ecclesiastical Province. The Diocese was carved out from the Owerri Diocese in 1994, with forty parishes and eight 'quasi-parishes'. The Anglicans have five Dioceses, with two new ones in the making as at the time of this research. In all there are 205 parishes in the five dioceses. The Methodists have two dioceses: the Port Harcourt Diocese and the Bori Diocese. The Port Harcourt Diocese has nine circuits and sixty four congregations, while the Bori Diocese has eight circuits and sixty five congregations. The Presbyterian Church of Nigeria in Rivers State has two Presbyteries – Port Harcourt and Port Harcourt North. The Port Harcourt Presbytery has five parishes, in addition to mission stations in Abonnema, Mbiama, and Bayelsa. The Port Harcourt North Presbytery has six parishes and at least two mission stations at Obuihe and Bodo City. The Baptists, under the umbrella of the Rivers Baptist Conference, have sixteen Associations and 306 local churches. The Pentecostal Fellowship of Nigeria, Rivers State Chapter, is divided into six 'Regions', each of which in turn is subdivided into 'zones'. There is a total of 107 zones spread across the state. Each zone has several local churches in its membership and most of the churches are autonomous and self-governing congregations of various shades and sizes.

The ubiquity of Christian witness in the region could be explained in a couple of ways. The first is its spread. There are virtually churches in the nooks and crannies of the major cities and towns as they vie for physical and sonic space. In addition to the spread of some of the mission-founded churches and the more widely known Pentecostal churches, such as the Redeemed Christian Church of God, The Living Faith Mission, alias Winners' Chapel, and Deeper Life Christian Ministry, others have sprung up in the region. Some of the key ones within Port Harcourt metropolis are Greater Evangelism World Crusade (founded by Apostle G.D. Numbere), Redemption Ministries (founded by Pastor Steven Akinola), The Royal House Of Grace (founded by Apostle Prince David Zilley Aggrey), Salvation Ministries (founded by David Ibiyeomie), The Royal Church (founded by Bishop Chinasa Nwosu), and Abundant Life Evangelical Mission (founded by Apostle Eugene Ogu, the Rivers State Chairman of PFN). Others are the Living Gospel Mission (founded by Rev. Fubara Ibama), Logos Ministries (founded by Bishop Yomi Isijola), Foundation Faith Ministries (founded by Bishop Sam Amaga), El Shaddai Church (founded by Bishop Elkanah Hansen) and The Hilltop International Christian Centre (founded by Pastor and Mrs. Chris Oahre).

A major feature of these churches and the second major pointer to Christian presence in the region is their competition for urban space and for new members through the electronic media and modern technology and through their imposing buildings, mass gatherings, and ocular adverts, such as massive billboards, banners and posters (de Witte 2008, 691; 2003, 172–202; Asamoah-Gyadu 2005, 336–57; Hackett 1998, 258-277.) The idea seems to be, 'the bigger and louder, the better and more attractive'. For example, the Rivers State Government-owned radio station, *Radio Rivers*, had at least thirteen different quarterly religious programmes sponsored by these churches in 2006, while the Rivers State Television, Port Harcourt *Channel 22* had a good number of religious programmes, also sponsored by these churches. Thus, according to its 2006 3rd Quarter Programme Schedule, the Television Station had at least twenty three of such programmes, including ones by the Hilltop International Christian Centre, Christ Embassy, Salvation Ministries, Redemption Ministries, and Greater Evangelism World Crusade. From the foregoing, it is obvious that Christian churches are well spread in the region.

## 1.3 Literature Review and Justification for Book

Several studies have been carried out on Nigeria and the Niger Delta vis-à-vis the oil industry and its impact on the region. These have led to several publications on oil in the country and region. Although it is difficult to appropriately categorise the studies since they address the various issues related to Nigeria's oil industry from several perspectives, they could reasonably be grouped for purposes of analysis as follows: political economy/foreign policy (e.g., Schatzl 1969; Pearson 1970; Onoh 1983; Shaw 1984; Ihonvbere and Shaw 1988; Ikein 1990; Khan 1994; Nnadozie 1995; Onosode 1998), human rights (e.g., Human Rights Watch 1999; Amnesty International, ERA/Friends of the Earth, and Okonta and Douglas, 2001), legal (e.g., Muhammed-Jallo 1989; and Ebeku 2002), conflict and oil-related litigation (e.g., Frynas 2000), policy and governance (e.g., Turner 1977; Brooks 2004), civil society (e.g., Ikelegbe 2001), and environment and health (e.g., ERA, Ebeku, and Steyn 2003).

From the above sketch, certain extrapolations could be drawn. The marked increase in the number of studies on the oil industry in Nigeria serves as an index to the world-wide attention it has received, especially in the light of the outcry against the alleged economic and environmental injustice brought about by the Nigerian Government and the oil companies on the region from where the oil is produced. The second point to note is that these studies are carried out by both Nigerian and non-Nigerian researchers and academics. And, thirdly, it is apparent that the economic issues in Nigeria's oil industry are in the front burner as they predominate in the studies. Individually and collectively the studies have been able to highlight the various intricacies associated with the oil economy as a whole and Nigeria in particular. They have helped to critique, complement, supplement, and/or expand the frontiers of research on Nigeria's oil economy within its socio-political and economic setting. In effect some of the studies complement each further.

For example, the two pioneering works on Nigeria's oil by L.H. Schatzl (1969) and Scott R. Pearson (1970) complement each other for, whereas Schatzl traces the economic history of Nigeria's petroleum industry, Pearson analyses and describes the impact of the industry on the country's economy up to 1968. Also while Manby's *The Price of Oil* (1999), which examines the issue from an individual rights' perspective, is complemented by Ebeku's thesis (2002), which looks at it from the perspective of the collective rights of the people of the region. A careful reading of the whole helps to understand the parts, and vice versa. The researchers have brought to bear on the studies their respective backgrounds, academic interests, professional goals, and intellectual expertise and as such have pushed the literature on the Niger Delta and Nigeria's oil economy to a level higher than it was in the 1970s.

The above positive remarks notwithstanding, a number of issues could be raised with these studies. With a few exceptions, most of them focus on the "superstructure" of the oil industry in Nigeria, namely, the political, legal and economic frameworks within which it operates, such as the enabling laws, regulations, and institutions that control, supervise and/or manage the oil-related agencies and companies in Nigeria and thereby downplay the "substructure," which is the people of the region from where the oil is produced. The above criticism will be illustrated with representative publications from each of the perspectives identified above.

Sarah A. Khan and T.E. Turner represent the political economy viewpoint. As the title of Khan's work indicates, *Nigeria: The Political Economy of Oil*, it is centred on the political economy of Nigeria's oil, especially on business operations, within the context of its history, economic policy and international relations. The book does not discuss how the oil affects the people of the region and/or the environment. Even in her discussion on the issue of gas flaring in Nigeria, vis-à-vis its impact on global warming, Khan is silent on the impact of such gas flares on the Niger Delta. The same criticism is applicable to Turner's study. Turner ably gives an in-depth analysis of the development of oil administration and policy in Nigeria between 1957 and 1977 and exposes the rivalry and bickering that existed between the Ministry of Mines and Power and the Nigerian National Oil Corporation (NNOC) before the two agencies of government were eventually merged to form the Nigerian National Petroleum Corporation (NNPC) in 1977. Due to the focus of the study there is no discussion on the impact of the industry on the people of the Niger Delta (See Turner 1977).

Ebeku's work represents the legal perspective as it examines the legal aspects of environmental and equity issues as they relate to the activities of the oil industry in the Niger Delta. He rightly argues that these activities are carried out as a collaborative venture between the Government of Nigeria and the MNOCs who receive huge revenue while despoiling the environmental, economic and social well-being of the people of the region. This neglect, he argues, has led to violent protests by the people of the region. A major asset of the thesis is the extensive discussion of the history of marginalization and fear of it among the ethnic minorities of the region. However, in his discussion on the unjust context within which the industry operates, he sees the injustice solely from the viewpoint of marginalization by the majority ethnic groups. Although the argument has some truth in it, it is not the whole truth as the author excludes other considerations such as corruption and the fact that the injustices are perpetrated by persons from several ethnic groups in Nigeria, including leading figures from the Niger Delta. This is amply illustrated by the allegations of corrupt enrichment and embezzlement of state funds by the impeached first civilian Governor of Bayelsa State, Chief Diepiriye Alamieyeseigha. Therefore, the scope of diagnosis of the problem and prognosis of the probable solutions should not be limited to ethnic or regional cleavages. Another criticism of Ebeku's otherwise well-researched work is his narrow definition of the Niger Delta, which he limits to the States of Bayelsa, Delta and Rivers. Geographically speaking, such a restricted definition may be tenable, but since the central point of the discussion and therefore the common denominator is oil, it is not reasonable to cut off other oil producing states from the analysis because the same deleterious effects from oil operations are applicable to all the communities where oil is produced.

Writing from an environmental point of view, Maria Steyn brought an innovative dimension into the research on the oil economy in Nigeria in her thesis. The title of the thesis is self-revealing: "Oil Politics in Ecuador and Nigeria: A Perspective from Environmental History on the Struggles between Ethnic Minority Groups, Multinational Oil Companies and National Governments." The study is a comparative analysis of the environmental and human impacts of oil developments on oil-producing ethnic minorities in Ecuador and Nigeria and the struggles of such ethnic minority groups against their national governments and the MNOCs. It uses the case study method and comparative analysis approach in examining the experience of the Cofan and Siona-Secova Indians in Ecuador and the Ogoni in Nigeria. Since the writer sees ethnic marginalisation as the sole paradigm to explain the oil-related injustice in both countries, especially Nigeria, without emphasizing the role of the "compradors" made up of persons of various ethnic groups, both majority and minority, it suffers from the same hermeneutical weakness as the work by Ebeku. According to Frynas, the term "comprador" is a Portuguese word for "buyer." He explains that "a comprador would perform the task of providing access to local markets for foreign firms." That is, he performs the role of a 'gatekeeper' or middleman for foreign firms (Frynas, 28, footnote # 47).

The work by Ike Okonta and Oronto Douglas is an excellent exposé on the travails of the people of the Niger Delta as a result of the tragedies brought upon them by the oil industry. Their study is essentially from a human rights perspective. In terms of content and analysis, the book does justice to its title, *Where Vultures Feast: Shell, Human Rights and Oil in the Niger Delta*. By singling out Shell for blame in the crisis, especially as it pertains to the Ogoni struggle, its case is weakened in scope because the crisis is beyond one company and, again, the Nigerian Government holds 55% equity in the Joint Venture Partnership (JVP) through the NNPC. The book also ignores the issue of how the Governments of the States and Local Government Areas have (mis)managed the extra revenue from oil in the form of the 13 per cent derivation fund since 1999. According to figures published by the Federal Ministry of Finance in Abuja, the nine oil producing States have received more oil revenue funds from the Federation Account since 1999 on account of the increase in the derivation fund and excess crude oil income. For example, records show that in 2005, these States received 36% of the ₦866 billion allotted to the 36 States of Nigeria.

Frynas' book, *Oil in Nigeria: Conflict and Litigation between Oil Companies and Village Communities*, dwells specifically on conflict and litigation in the region. Using a socio-legal approach, Frynas analyses the conflicts between the oil companies and the oil-producing communities, which he calls "village communities," by focusing on litigation. He gives a helpful perspective on the oil-related conflict in the region but the book suffers from two major weaknesses. Whereas the stated focus of the study as could be deduced from the title is 'conflict and litigation', the book does not dwell sufficiently on the issue of conflict; rather it pays more attention to litigation. Also, in the gathering of data for his analysis, Frynas failed to interact with the people affected by the oil operations. Instead, he draws his information from a survey of 154 Nigerian lawyers; thus the voice of the real sufferers is not heard in the study.

A unique and very useful study deserves some mention. It is an article by Augustine Ikelegbe in which he carries out an elaborate discussion of the oil economy as it pertains to the Niger Delta from the perspective of civil society. His study is a broad description of the development, types, profiles, methods and ramifications of the various civil society groups in the Niger Delta vis-à-vis the quest for 'resource control'. He concludes from his study that "civil groups have reconstructed the agitation into a broad, participatory, highly mobilised and coordinated struggle and redirected it into a struggle for self-determination, equity and civil and environmental rights" (Ikelegbe, 437).

So far, it is apparent from the above review that no academic study has been carried out on the issue of oil and conflict in Nigeria from the perspective of Christian social witness. Two studies merely make reference to the church. Augustine Ikein uses a Methodist view of systems theory to argue for responsible behaviour that places in proper balance economic benefit and human ecology in his discussion on the need to care for the oil-producing areas. L. J. Brooks includes religious associations and church groups in her discussion on local and international NGOs and other groups. She argues that whereas local NGOs suffered under the military in Nigeria, church groups flourished. She therefore avers that these church groups could have served as veritable instruments in the delivery of community needs, but according to her, they failed to utilize their vantage position.

To the best of our knowledge, this book is the first study of Nigeria's oil industry from the perspective of Christian social witness. In doing so, it adopts approaches in practical theology, especially the emergent field of congregational studies, with its emphasis on qualitative research, and African Christian Theology. There is also increasing emphasis on grassroots theology, or theology from below, especially the approach that deals with the socio-political challenges of the continent that runs through such African theologians as Jean-Marc Ela (1986, 1988), Jesse Mugambi (1995), Kä Mana (2004), and Charles Villa-Vicencio (1992). Whereas the operative word in Ela's analysis is *liberation*; that of the last three is *reconstruction.* Also, whereas the main emphasis of Ela's theology is on human liberation with focus on praxis (Parratt 1995, 142), the theology of reconstruction takes it further by rejecting the tendency to polarise the themes of *inculturation* and *liberation* and by integrating them and insisting that Jesus engaged in issues of personal liberation and social transformation in his ministry. However, it has been argued elsewhere that even the theology of reconstruction is not broad enough as it does not sufficiently grapple with matters of the non-human environment (Osuigwe 2012).

## 1.4 Research Questions, Objectives and Scope

The pivotal questions this book seeks to address are:

1. What are the underlying theological issues and beliefs of the three congregations in this research within the context of oil and conflict? Of particular interest here in the light of the allegation that African evangelical Christianity is preoccupied with soul winning is an investigation and analysis of their theology of conversion.
2. What is their prevalent praxis within the context of Nigeria's political economy of oil and conflict?
3. How accurate is the view represented by Paul Gifford that African evangelical and 'fundamentalist' Christianity lacks social responsibility and is a-political and anti-development?
4. How would a contextual political theology for Nigeria's political economy of oil look like?

The above questions help to delimit the main objectives of this study. The objectives are first, to investigate, document, analyse and clarify the theological and social consciousness of Baptist and Pentecostal Christians in Nigeria, within the context of the Niger Delta conflict, and second, to use the data to examine the veracity or otherwise of the claims that African evangelical Christianity is so other-worldly in orientation that it lacks this-worldly responsibility. There are also two ancillary objectives. The first, which seeks to provide the socio-political background and facilitate the achievement of the main objectives above, is an examination and critique of Nigeria's political economy of oil and conflict,while the second is to propose and sketch a theological perspective to the conflict in the form of a political theology.

The scope and limit of this book are circumscribed by the above objectives. Thus, no attempt is made to address the views and perceptions of other mission churches, such as the Roman Catholic, Anglican, Presbyterian, and Methodist Churches. It is also outside the scope of the book to examine the views of the African Indigenous Churches (AICs). Rather the study focuses on the witness of two local Baptist congregations and a Pentecostal church in the region with particular reference to the question of oil and conflict.

## 1.5 Methodology

This book is divided into two major parts, each of which requires its own approach in data gathering and analysis. The first part, which lays the foundation for the second, depends on both primary and secondary sources for its data gathering and analysis. It is covered in the critical review and analysis of the Gifford literature and the literature on Gifford that follows in Chapter Two and the discussion on Nigeria's political economy of oil and conflict in Chapter Three.

The second part, which is on the investigation of Christian witness in the Niger Delta within the context of oil and violence, depends mainly on primary sources through an ethnographic field study using the case study approach. Thus data was gathered using focus group discussion, interviews, participant observation, and spontaneous conversations. Also, members' prayers, songs and sermons were noted and analysed. For the case studies, the following churches were selected: First Baptist Church, Buguma City in Asaritoru Local Government Area of Rivers State ('First Baptist' henceforth), Faith Baptist Church, Port Harcourt (hereafter, 'Faith Baptist'), and The Hilltop International Christian Centre, Port Harcourt ('Hilltop' henceforth).

There are three major reasons for their selection: theological, geographical, and pragmatic. The key theological ground for the choice of these churches is the quest to test out the 'received account' on African evangelical and 'fundamentalist' Christianity regarding social witness. This is aptly represented by Paul Gifford who alleges that the only thing that matters to such churches is evangelism, and that they lack social responsibility and commitment to development, since "only Christians matter" (Gifford 1991, 9-20; Odion-Akhaine 2006, 758-760). Hence this study sets out to investigate the correctness, or otherwise, of such claims using data from the case studies.

The location of the churches is germane to their selection. Port Harcourt is at the heart of Nigeria's hydrocarbon industry and the leading city in the Niger Delta in terms of its economic clout and political importance. Also, it has had its own share of the violence in the region which has impacted on its inhabitants, who come from the region and beyond. For instance, the membership of Faith Baptist and Hilltop cuts across the various ethnic, social, economic and political strata in the country. Since they live in the City, they have had direct impact of the conflict. It was therefore felt that interacting with them would afford this author the opportunity to get the various perspectives from the urban grassroots as much as possible. First Baptist was chosen because it is the oldest Baptist Church in the region and has had several experiences of the violence. These are in addition to the fact that it enabled the researcher to get the rural/semi-urban viewpoint. The choice of Hilltop is intended to serve the purpose of gauging the Pentecostal perspective in the analysis. It is a medium-sized Pentecostal church founded and led by Pastor and Dr Mrs. Chris Oarhe.

The pragmatic ground of the choice of the churches is simply accessibility within the time-limit of the study. Being a Baptist, getting the confidence and approval of the two Baptist Churches was fairly easy. Hilltop was not the Pentecostal church initially earmarked for the research. Originally it was the Royal House of Grace, Port Harcourt but, when they were approached, they declined. For this reason, the researcher contacted and sought the approval of the Senior Pastor of Hilltop, and he obliged.

However, before outlining the demographic data on the field research, it is pertinent to highlight the difficulties encountered during the course of field work carried out in July to December 2006 and July to September 2007, spanning a period of eight months. The research was conducted under very difficult and dangerous circumstances as a result of unbridled violence that engulfed most of Rivers State in 2006 and 2007. The period was marked by inter-gang warfare between militant youth groups, on the one hand, and fierce battles between the Joint Task Force, made up of members of the Nigeria Police and the armed forces, and the various gangs in the state, on the other. Put simply, Port Harcourt was like a war zone during most of the period. Thus, social and economic activities were grounded as movement of people and goods within Port Harcourt and its environs was highly restricted. Schools and offices either were closed in the course of the day or were completely shut for some of the period.

Matters were made worse by the sporadic and unpredictable nature of the battles as they could erupt suddenly from any part of the city and when they did, they lasted for an entire day in some instances. The situation affected the programmes and activities of the three churches and in effect impacted on the research plans as the situation forced an adjustment to the original number and variety of focus group sessions. Also, due to safety concerns, this researcher could not move around as much as he intended. The number of focus group and interview sessions in the congregations was therefore circumscribed by two inter-related factors, namely, the respective calendars of the three congregations and the prevalent condition of gang warfare which led to cancellations of several of their programmes. Indeed the courage of members of these churches to attend and participate in the research during the period of fieldwork in 2006 and 2007, despite the gun battles, is commendable and is highly appreciated. The dangers of the period could be highlighted by the death on August 6, 2007 of Mrs. Margaret Ibama, who was a member of Faith Baptist, through the stray bullets of militants, while on her way back from work. Despite all this, all necessary steps were taken to ensure that no member of the churches was unduly put at risk in the course of the research. Some of the steps include the postponement of scheduled meetings where reports indicated ongoing gun battle in particular areas and careful timing of the meeting so that the subjects of the study could safely get to the venues and return home thereafter. Reports of gun battles came either through the electronic media and/or phone calls and text messages which members sent to one another.

Thus, in all seven focus group discussions were conducted in the three churches: two each in First Baptist and Hilltop, and three in Faith Baptist. Originally the plan was to hold several small group sessions with a cross-section of members of the three churches; but, due to the volatile nature of the period occasioned by gang warfare, enlarged sessions were held with them in the evenings on weekdays when the churches normally hold their Bible Studies and/or prayers. The two focus group sessions in First Baptist took place on Thursday August 31, 2006 and October 12, 2006, and each lasted for two hours on average. There was on average 109 members in attendance at the sessions. The breakdown of attendance at the meeting of August 31, 2006 is as follows: thirteen adult men, forty four adult women, twenty six youths, and twenty five students. Out of these, thirty one persons were male while seventy seven were female. The number of ethnic groups represented at the focus group sessions in the church were thirteen and they were mainly drawn from the South-Southern, the South-Eastern, and the South-Western geo-political zones of Nigeria. Nigeria has six geo-political zones, three in the North and three in the South. The three zones mentioned here are in the South. The northern zones are North-West, North-Central, and North-East.

The two focus group sessions in Hilltop recorded the highest attendance among the three congregations. With the understanding of the church leadership, two three-hour sessions were held with the church on Wednesday September 6 and October 4, 2006, respectively, after several failed attempts at finding dates that were convenient and safe. In attendance at the meeting of October 4, 2006 were 290, including 29 adult men, 40 adult women, and 221 youths and students drawn from twenty different ethnic groups in Nigeria. The male/female ratio at the meeting was roughly 50:50. These included fifteen persons with master's degrees and above, 126 Bachelor's degree holders, fifty eight Secondary School leavers, and six with the First School Leaving Certificate.

The same difficulties encountered in fixing dates for the focus group discussions in both First Baptist and Hilltop were encountered in Faith Baptist. Eventually two church-wide sessions were held in the church on Wednesday August 2 and September 13, 2006. In addition to the overall situation of violence in the city, the two meetings were further affected by two respective additional factors: a twenty four-hour heavy downpour that led to profound flooding in Port Harcourt and a nation-wide strike embarked upon by the two leading unions of oil workers in Nigeria – PENGASSAN and NUPENG, which further paralysed movements. A separate session was held with twenty nine members of the Royal Ambassadors (an organisation for boys within the ages of ten and twenty five) of the church on August 11, 2007 after three failed attempts as a result of the violence in Port Harcourt. The total attendance at the church-wide focus group discussion held on August 2, 2006 was 153, with females forming the majority. Among them were 36 adult men, 39 adult women, 48 youths and 30 students. They included six persons with postgraduate qualifications, thirty eight first degree holders, twenty post-primary school leavers, and six persons with primary school qualification. Their ethnic composition was twenty one. The proceedings of the focus group sessions, which were conducted mainly in English, were recorded both electronically (on tape) and manually. Those that assisted with the manual recording in Faith Baptist include Prince Nyekwere, a lawyer, and two students, namely, Sotonye Igoniko and Prince Ekine; in First Baptist they were three students: Krakafa-a Briggs, Mondayba Philip, and Ipalibo G. Princewill; while Azuatalam Osuigwe, a chartered accountant, took the records in Hilltop.

The focus group sessions were complemented with participant observation, personal interviews, spontaneous private conversations with individuals and/or groups, phone calls, and further correspondence for the clarification of issues encountered in the course of data analysis and interpretation. The interviews, which were of three types – the schedule-structured interview, the focused interview, and the non-directive interview (cf. Frankfort-Nachmias 1977, 541-557) – were conducted using open-ended questions so as to gather as much insight as possible and to document and describe their "*emic*" perspective to the conflict. "*Emic"* perspective refers to the insider's perspective, or the "native's point(s) of view," and is usually gained through participant observation and interviews based on open-ended questions, in addition to site documents. Its opposite viewpoint is the "*etic,"* or "outsider," perspective." This was intended to help gauge the perspective of the churches in three key areas: their experience of the conflict, their understanding and interpretation of the experience in the light of their faith as Christians, and their response and witness in the light of that interpretation (cf. Boff 2000, 198).

Through these means additional direct contacts were made with eleven members of First Baptist, ninety seven of Faith Baptist, and fifty two of Hilltop, making a total number of 120, 250 and 343, respectively, in the three congregations. They include kidnap victims, a number of former militants, and persons who initially considered joining the armed struggle as an option. The educational qualifications of the interviewees ranged from none at all to PhDs. Among them were pastors; oil company staff; civil servants; academics; professionals – architects, medical doctors, lawyers, engineers, nurses, teachers, scientists; private businessmen/women and petty traders; students; and unemployed persons.

The settings in which the churches were observed include their worship sessions, Bible Studies, prayer meetings, and business meetings. For example, First Baptist was observed on Sunday August 27, 2006 and Sunday August 12, 2007, during their regular worship, and on Thursday August 31, 2006 during their Bible Study session that preceded the focus group discussion; while Faith Baptist was observed on at least three different occasions. The highlight of the details of the observations is given in Chapter Four below, but suffice it to state here that Hilltop with a membership of 1500 has an average Sunday worship attendance of 1200; Faith Baptist with 1200 members records on average 1100 on Sundays in its two worship services; while First Baptist with about 960 members has an average weekly attendance of 635.

Drawing insights from grounded theory (cf. Glaser and Strauss 1980, 2), the strategies of data analysis utilised were transcript-based, note-based and coding (Krueger 1994, 143-144) as determined by the rules of clarity, parsimony, logical coherence, density, scope and integration (see Glaser and Strauss). The units of analysis used include the statements, activities, and the individual and corporate experiences of the three churches. That is, data was collected manually (through note-taking) and electronically (through tape-recording). They were then transcribed, analysed and coded. The analysis and coding involved an ongoing process of finding and integrating recurring themes in the statements, sermons, prayers, and songs on the basis of their frequency and distribution (cf. Janesick 1982, 19). Due to reasons of space and focus, the minimum number of occurrence for each category is an aggregate of ten distributed across the three congregations.

Consequently, the process yielded the following major categories: the Church, Jesus Christ, God, the Holy Spirit, Prayer, Conversion (evangelism and soul winning), Crude Oil, Conflict, Violence/Non-violence, Resource Control, Militants, Christian Social Involvement, Oloibiri, and Marginalisation. To facilitate and enhance the analysis, these are further grouped under three headings: The Trinity – God, Jesus Christ, and the Holy Spirit; The Church, Meaning, Purpose and Social Role; and Crude Oil and Conflict – crude oil, violence and non-violence, militants, resource control, Oloibiri, and marginalisation. The conveying of the outcome of the research involves the use of narrative, description, critical analysis and evaluation. Because it is intrinsically connected to their beliefs, the theme of prayer is subsumed under the doctrine of God.

The methodology outlined above has some pedigree in theological scholarship, as will be illustrated with some examples. The first is Nancy Tatom Ammerman's *Congregation and Community*. The book is a study of twenty three local congregations selected from nine different communities experiencing social change in the US. In each of the churches studied, Ammerman investigates three major dimensions, namely, their resources, structures of authority, and culture. Data for the study were gathered through participant observation, interviews with church leaders, staff, and some members, and by a survey of some congregants. The central argument of the book is that local churches have two options in the face of changing environment: adapt or face decline. She rightly argues that congregations are "a fertile source of social capital and a powerful instrument of social integration" (Wilson 1997, 769). In all, the book is rightly adjudged to be "extremely useful to those engaged in congregational research" (Kanieski 1997, 188).

However, this study differs from Ammerman's in at least four major areas: the context of the research, the number and types of congregations investigated, the dimensions investigated, and the main theme examined. Whereas Ammerman's context of study is in the United States, this study focuses on local congregations in Nigeria; and while the number and types of congregations investigated by Ammerman are more extensive (twenty-three), those used in this study are three. Moreover, whereas Ammerman looks at the dimensions of resources, structures of authority and congregational culture of each of the churches and has social change and the churches' response to it as her central issue of investigation, this work is concerned with investigating Christian witness within the context of oil and conflict in the Niger Delta region of Nigeria, and using the data to evaluate the received account on African evangelical and 'fundamentalist' Christianity.

The second example is Don S. Browning's *A Fundamental Practical Theology: Descriptive and Strategic Proposals*. The driving question that Browning grapples with in the book is on how local congregations and religious communities make sense and in what ways they are both "communities of memory" and "communities of practical reason" (Browning 1991, 3). It identifies with and seeks to contribute to the emerging field of congregational studies spearheaded in the US by the Lilly Endowment, which is an Indianapolis-based, private philanthropic foundation that seeks to strengthen Christian congregations by providing an opportunity for pastors to retreat a while away from their work so as to engage in a period of renewal and reflection. Browning asserts that through such studies congregations are carefully and critically analysed historically, sociologically and theologically (Browning 1991, 2).

There are two significant ethnographic studies on African theology/Christianity that were quite helpful in the choice of approach for this book. The first is by Gifford. It is the 1987 journal article, "'Africa Shall Be Saved': An Appraisal of Reinhard Bonnke's Pan-African Crusade" (Gifford 1987, 63-92). The paper is mainly an effort by Gifford to systematically establish Reinhard Bonnke's Christ for All Nations (CfAN) "brand of Christianity" by analysing the thirteen sermons that were preached by a number of pastors at a thirteen-day crusade held in Harare, Zimbabwe in 1986 by CfAN. The article gives a first-hand description of the setting of the crusade and then analyses the sermons under nine headings: miracles, The Bible, Christ, the Spirit, demonology, the sacraments, eschatology, ecclesiology, and morality. The analysis led him to conclude that CfAN is squarely a Pentecostal group with links to "the revivalist, Pentecostal, fundamentalist Christianity of the southern states of America" (Gifford 1987, 85).

The second significant ethnographic study of African Christianity that influenced the approach in this study is the excellent examination of current Christological trends in Africa by Diane Stinton. The book, which is based on her 2001 PhD thesis at the University of Edinburgh, utilises a methodology that "integrates the analysis of written or textual Christologies with that of oral Christologies gained through qualitative field research." Continuing, she adds that a "qualitative approach serves theological analysis especially well, since it focuses on '*participant perspectives*' and seeks to discover 'what *they* are experiencing (and) how *they* interpret their experiences" (Stinton 2006, 18, emphasis in the original). This is exactly the case with this work as it seeks to convey and interpret the *emic* perspectives of the participants in the case studies. But the approach here differs from Stinton's in terms of sources. Whereas her study combines data from three sources – African Christological texts, the authors of those texts, and African Christians from various church traditions in Kenya, Uganda and Ghana – this book draws its sources from grassroots evangelical and Pentecostal Christians in three local congregations in Nigeria as a result of its focus, especially their oral theology, as they deal with a socio-political crisis, namely, the conflict centred on oil.

Since congregations are the sum total of their members such that the corporate and individual aspects are inseparable, attempt is made in the case studies to distinguish and as well integrate the individual and corporate views of the three churches. Individual opinions are gauged through responses to questions during the focus group sessions and from private interviews, discussions and conversations; while collective views were accessed from official church documents where they are available. Some of the documents include the Statements of Faith and Practice, History, Constitution and Bye-Laws, and data placed on their websites.

The need to grapple with African grassroots theology and, therefore, reverse the normal theological methodological order (Peter 1994, 107), has been noted by a number of scholars. John Mbiti, for example, commends it in his delineation of three approaches to theologising in Africa, namely, written, or academic, theology; oral, or grassroots, theology; and symbolic theology that focuses on art forms (Mbiti 1979). He also states:

> Much of the theological activity in Christian Africa is being done as oral theology… from the living experiences of Christians. It is theology in the open, from the pulpit, in the market-place, in the home as people pray or

> read and discuss the Scriptures...African Christianity cannot wait for written theology to keep pace with it... Academic theology can only come afterwards and examine the features retrospectively in order to understand them (Mbiti 1986, 229).

The preceding three approaches correspond roughly to Charles Nyamiti's three schools of African theology. Respectively, they are the speculative, the socio-biblical, and the reactionary (Nyamiti 1973, 1; Peter 1994, 2). For long African theology, under the influence of Western theology (Peter 1994, 106; Pailin 1986, 123), was dominated by written theology, or the speculative school. But emphasis has shifted towards grassroots theology, or theology from below (Goergen 2007; Parratt 1995, vii), in which theology is not solely conceived as "a theology of theologians, by theologians, and for theologians," but as "a theology of the people, by the people, and for the people" (Peter 1994, 105); or as theology done "*in the context of* and *with* the people" (Stinton 2005, 108). Cameroonian theologian Jean-Marc Ela calls this collaborative, communal approach to theology as "shade-tree theology" (Ela 1986, vi; Ela 1988, 11); while Adrian Hastings calls it "implicit theology" (Hastings 1976, 54). Some examples of such people theologies that exist in other parts of the world include the *Minjung* theology in Korea and "the theology of struggle" in the Philippines.

The above viewpoint is further supported by James C. Okoye who argues that there are two parts of African Christian reality that deserve greater attention in the African theological enterprise, namely, "the 'oral theology' buried in sermons, songs, and popular literature and the 'practical theology' of the people as they strive to live the faith in context" (Okoye 2007; cf. Burgess 2008, 30). In the context of Nigeria, Catholic theologian John Onaiyekan describes some strands of this grassroots theology as "silent theology," "instant theology," and "gossip theology" (See Kukah 1995, 228-229).

In this transition in approach, the primary task of academic theology, according to C.B. Peter and Donald Goergen, is to "understand and scientifically interpret how people are doing it and what are its implications" (Peter 1994, 107), since African people are already doing theology (Dickson 1984, 13), even if they are not aware of it, or "to be in contact with the people's theologies in order to discern what the Spirit is doing in the churches" (cf. Goergen 2007). Also, Kwame Bediako rightly argues that without this vital link with grassroots theology, "academic theology anywhere can become detached from the community of faith and so be not much more than an exclusive conversation carried on among the guild of scholars, and incapable of communicating life in Jesus Christ to others" (Bediako 2004, 17-18; Bediako 1993, 15).

This approach has some inherent advantages: (1) it treats African Christians as subjects rather than objects of study as it underscores "the importance of the community of faith in theological formulation" (Stinton 2005, 108); (2) it helps bridge the gap between academic theology and grassroots theology by ensuring that African Christianity is not turned into a solely intellectual enterprise without practical relevance or direct bearing on the real life challenges being faced by Africans, since it "comes from where the faith lives and must live continually, in the conditions of life of the community of faith..." (Bediako 1993, 5); and consequently, (3) it has the capacity to save 'professional' theologians from the error of answering questions that nobody is asking. For example, in his critique of Western theology, Desmond Tutu has pointed out that it has "some splendid answers, but they are answers to questions that no one elsewhere is asking!" (See Parratt 2005, 8). The above three advantages are exploited in this book on account of the methodology outlined above.

Consequently, the research findings from grassroots theology perspective will be used to test the validity, or otherwise, of the claims of academic theology, as represented by Gifford, regarding evangelical and 'fundamentalist' Christianity, against the backdrop of Nigeria's political economy of oil and conflict drawn from relevant texts. Also, the book will interact with the views of academic theology in the course of the analysis. This is especially the case in the discussion on Christology in chapter five. The specific claims to be examined in the research are outlined in the next chapter.

## 1.6 Format

As indicated above, this book has two major parts. The first part covers chapters one to three and it lays the foundation and sets the stage for the second part, which is covered in chapters four through eight. Chapter one introduces the study, while chapters two and three respectively set the theological and political contexts of the study by discussing the Gifford literature/the literature on Gifford and describing and critiquing Nigeria's political economy of oil. Chapter four gives a detailed ethnographic description of the three churches in the study. Chapters five, six and seven dwell on the documentation, description, and analysis of the data from the three churches. While chapters five and six focus on the theological themes of the Trinity and the church, respectively, chapter seven is on crude oil and conflict. The last chapter evaluates Gifford's claims on

African evangelical Christianity on the basis of the findings, raises some implications and challenges arising from the study, and recommends some issues that require further research in the light of the findings. It will also propose a political theology for Nigeria's oil economy in the light of the study.

## 1.7 Clarification of Terms

There are a number of terms used in this book whose meaning require clarification. They include "cults," "cult groups," "youth," "youth groups," "adult," "student," "evangelical," and "fundamentalist." The term "cults," or "cult groups," in Nigerian usage refers to a number of secretive groups with semi-religious attributes that originated among Nigerian university students, beginning with the first one that was established in 1952 at the University of Ibadan. Most of the groups later evolved from benign campus organisations into violent criminal gangs that now operate branches in local communities and urban centres in order to bolster their status and operations (Human Rights Watch 2008, 17). Each cult group usually has its own initiation rites, paraphernalia, drinks, bodily marks and area of dominance. It is believed that several leading politicians belong to some of the cult groups. A number of such groups are highlighted in chapters three and seven.

By "youths" is meant those between the ages of sixteen and thirty five; while "youth groups" refers to a motley of associations and organisations in the Niger Delta organised under various umbrellas, either as the youth wing of ethnic organisations and/or of community development unions, or as peer fraternities for the promotion and protection of group interests. Such group interests may be economic, social, and/or political. Some youth groups serve as vigilantes organised and used by politicians to further their political goals. Some of these youth groups are also mentioned in chapters three and seven. 'Adult' is used in this book in reference to men and women above the age of thirty five; while 'students' refers to those currently schooling at either the post-primary or post-secondary level.

The term "evangelical" is derived from the Greek word "*euangelismos*," "gospel" or "good news," which was used by Martin Luther during the Reformation to describe his movement as the *evangelische kirke* (evangelical church). Later in Europe "*Evangelical*" became a synonym for "*Protestant*," a meaning which it retains in Germany today (Marshall, 2005, 22-23). In North America, there is a lack of unanimity on its meaning. Some definitions given to it in the region include: “a specific conservative Christian system of beliefs,” “a religious experience,” “a commitment to a proselytizing activity,” “a style of religious service,” “a walk with God,” or “a group of denominations” (Marshall 2005, 22-23). However, “evangelical” is used in this book with reference to the form of Christianity that affirms the necessity of personal conversion, the propriety and imperative of evangelism, the inspiration and authority of Scripture, the bodily resurrection of Christ and his personal return. The term ‘fundamentalist’ in the contemporary world connotes a number of things to different people and generates quite a number of emotive responses as a result of its association with extreme forms of religious fanaticism and bigotry (Martin, 1996, 5-12). Despite some reservations about the term (more will be said on this in chapter two below), it is used for purposes of analysis to refer to the form of Christianity that holds a conservative view of the Bible and morality and which tends to dispensational millenarianism. More discussion on “evangelical” and “fundamentalist” as they apply to African Christianity is given in the next chapter.

# CHAPTER TWO

# PAUL GIFFORD ON AFRICAN CHRISTIANITY

## 2.1 Introduction

This chapter sets out to review Paul Gifford's assessment of African Christianity, especially on the attitude of evangelical and "fundamentalist" churches towards social witness. It will also evaluate how he is being viewed among scholars of African Christianity. Gifford was once a lecturer at the Universities of Zimbabwe and Leeds. Before his retirement, he also taught at the Department for the Study of Religions of the School of Oriental and African Studies (SOAS), the University of London. His particular area of research is on present-day Christianity in Africa and its public role (Gifford 2004, 176). The aim is to set the stage for the case studies whose outcome will be used to evaluate his claims regarding African evangelical Christian social witness.

The main reason for the choice of Gifford in this study, in addition to the limitation of space, is his leadership position on the subject matter, which could be explained in four inter-related ways. The first, as will be seen from the literature review below, is the fact that he has published extensively over a period of time on African Christianity, generally, and on the evangelical and Pentecostal brand of that Christianity, in particular. Second, and on account of the above, he has influenced a number of African and non-African scholars on African Christianity and/or elicited responses from others. Such responses are classified and analysed below. The third related ground of focus on Gifford and which derives from the first two above is peer affirmation as another major scholar on African Pentecostalism, Allan Anderson, has described him as "a leading exponent on this subject" (Anderson 2007, 9), or "the leading and voluminous exponent of this subject" (Anderson 2002, 4). Fourth, no other scholar of African Christianity has made such a set of claims on its evangelical expression over such a period of time as he has done. Added to the above is the fact that, hitherto, there has not been any deliberate attempt to empirically test out his claims on African evangelical Christianity regarding social witness. The literature review that follows will also bear testimony to this.

## 2.2 Gifford on African Christianity

There are two major classifications of African Christianity by Gifford. The first is contained in *Christianity and Politics in Doe's Liberia*. Following an overview of Liberian history, Gifford delineates four categories of churches in the country: the mainline churches, the evangelical churches, the faith gospel, or "health and wealth," churches, and the independent churches. The second classification is found in *Ghana's New Christianity: Pentecostalism in a Globalising African Economy*, in which he identifies five strands of Christianity in Ghana in 1980: the Catholics, who were the largest; the mainline Protestant churches – the Methodist, the Presbyterian, and the Anglican; the established Pentecostals (Apostolic Church, Church of Pentecost, Christ Apostolic Church, and Assemblies of God); the African Independent (or Initiated) Churches (AICs), and the 'New Christianity'.

The difference in the two classifications above highlights the difficulty in putting African Christianity into straitjackets, especially in the light of the growth and diversity of the Pentecostal movement in the continent. Anderson succinctly captures this when he says, "The entrance and pervading influence of many different kinds of Pentecostalism on the African Christian scene now makes it even more difficult, if not impossible, to put African churches into types and categories" (Anderson 2002, 2). This is partly because, as Akintunde Akinade remarks, "African Christianity is remarkably protean, always changing and always in a process of renewal" (Akinade 2006, 15). Thus, any attempt at integrating them would yield minimal results as it is "fraught with great difficulties and requires continual rethinking and flexibility" (Burgess 2006, 329). For example, whereas both the Catholics and the mainline Protestant churches in the Ghanaian classification may fall under the mainline churches grouping in the Liberian setting, and the faith gospel churches in the first classification correspond to the New Christianity in the second, the evangelical churches (in the case of Liberia) do not necessarily correspond to the established Pentecostal churches (in the case of Ghana). Moreover, Gifford's classification is limited to churches south of the Sahara.

This weakness is somehow remedied by Katherine Marshall, who offers a broader classification that includes northern Africa. She delineates African churches as follows: the Roman Catholic Church; traditional Protestant Churches, which mostly originated from $19^{th}$ missionary activities; Orthodox Christianity, represented by the Ethiopian Orthodox Church; African Independent/Indigenous Churches (AICs), many of which have at least some ties to traditional practice; and Evangelical/Pentecostal churches (Marshall 2005, 3-4). Yet, her grouping is fraught with difficulties, one of which is the fact that there are charismatics within the Roman Catholic Church, while a number of Protestants are both evangelical and Pentecostal. She acknowledges this difficulty when she notes that several boundaries are "traced among them" (Marshall 2005, 3).

Therefore, the most helpful classification for the purposes of this book is the one that is based on Christian social witness, which coalesces the above categories into two broad groups. According to Gifford, the two broad groups are those that engage with the social deterioration in the continent and those that either deflect attention from it or fail to confront it. He gives positive assessment of the contribution of the former towards Africa's socio-political development and promotion of civil society, particularly from the 1980s. According to him, the churches' role in the wider society during the period went beyond the traditional sectors, such as education, health and development, to include direct political involvement. The last role includes challenges to autocratic political structures, urging reform, advocating political change, and even presiding over national conferences intended to midwife reforms.

Some of the countries where church leaders presided over national conferences for political reform include: Benin, where Mgr Isidore de Sousa, Archbishop of Cotonu presided; Gabon, where Mgr Basile Mve Engone, Bishop of Oyem presided; the Congo, it was Mgr Ernest Kombo, the country's leading Jesuit and Bishop of Owambo; and Zaire, where it was Mgr Laurent Monsengwo Pasinys, Archbishop of Kisangani (Gifford 1995, 3-5; cf. Gifford 1998, 21; and Haynes 2004, 71). By playing such roles, he argues, the churches demonstrated that their power in the political economy of their respective nations goes beyond the spiritual to include the political, the moral and the social. However, he notes that unfortunately and paradoxically, some churches played key roles in opposing reforms in some countries by supporting oppressive regimes. This was the case in Kenya under Arap Moi, and in Togo, Malawi and Zambia (Gifford 1995, 3-5; cf. Ojo 2004, 1). Bearing the above classifications in mind, it is pertinent to review some of Gifford's key publications on African Christianity.

## 2.3 Review of Key Gifford Publications on African Evangelical and "Fundamentalist" Christianity

Gifford has written extensively on African Christianity, either as author or editor. However, for purposes of this book, the focus here is on six of his works, as follows: *Christianity: To Save or Enslave?* (1990); "Christian Fundamentalism and Development" (1991); *Christianity and Politics in Doe's Liberia* (1993); *African Christianity: Its Public Role* (1998, 2001); "The Complex Provenance of Some Elements of African Pentecostal Theology" (2001); and *Ghana's New Christianity* (2004). This is with particular reference to his assertions on evangelical and "fundamentalist" churches in Africa. These studies will be examined chronologically.

The first publication to be reviewed is the twenty-five page booklet, *Christianity: To Save or Enslave?* In it Gifford looks at the spread of Christianity in Africa in the 1980s, a period tagged "a lost decade" for Africa, within the context of the economic and social decline that occurred during the period. His major arguments in the book are (1) that Christianity in the continent should grapple with the social, political, and economic malaise confronting the continent; (2) it should not trivialise such problems; and (3) it should not succumb to or aggravate them. His aim is to demonstrate that Christianity comes in various forms which aggregate to two distinct sets – those that engage with the social deterioration and those that either deflect attention from it or fail to confront it. Using his scheme as a basis, 'fundamentalist' churches belong to the latter.

"Christian Fundamentalism and Development" is basically Gifford's examination of the key features of the "fundamentalist" form of Christianity which emerged in early 20th century America and re-emerged as a significant trend in the late 1970s. He describes the spread of this form of Christianity to Africa in the 1980s and its implications in the present crisis in Africa. His main argument is that these features, all or some of which are found in particular 'fundamentalist' theology, encourage a passive acceptance of disasters, misfortunes and a lack of social responsibility, leading to the absence of any commitment to development, anti-socialism; and Christian Zionism that accords the nation of Israel "blind" support, which he claims is as a result of American influence. He had earlier made similar claims in *Christianity: To Save or Enslave?* in which he says that Christianity comes in various forms that aggregate to two distinct sets – those that engage with socio-political issues and those that either deflect attention from them or fail to confront them. For example, in his discussion on millenarian dispensationalist churches, he alleges that their preoccupation with the end-time and the return of Christ leads them to three tendencies: passive acceptance of social ills as part of the outworking of God's will and as precursors to the end-time, since "nothing else matters but preaching the gospel, or fulfilling the Great Commission (Mat 28, 19)." Moreover, he states, "Not only does this divert attention from the social, economic and political realities of most Africans, by teaching about them; it teaches that there is something almost unchristian about worrying too much about them" (Gifford 1990, 4).

In *Christianity and Politics in Doe's Liberia*, Gifford examines the exponential growth and role of Christianity in Liberia during the 1980s, which was a period marked by widespread oppression and mass failure in economic and social life. He analyses the support for oppression and the status quo that the growth paradoxically encouraged. He also evaluates the role of American denominations within the context of US foreign and economic policy during the period. Of particular interest is his view on evangelical churches which he defines as those churches that preach what he calls a "Liberian theology," a theology that lays emphasis on allegiance to government, spiritual combat between good and evil spirits, and biblical fatalism that leads to an anti-political attitude towards the society (See Sullivan 1995, 368-371).

Perhaps his most influential and well-known book is *African Christianity: Its Public Role*, in which he adapts ideas from political economy to describe contemporary manifestations of African churches and churchgoers, and appraises their various contributions to political developments in the continent since 1989. He also examines the roles of the churches in promoting civil society in Africa (cf. Ward 1999, 272). The African countries he uses for his case studies are Ghana, Uganda, Zambia and Cameroon. Following an overview of Christian churches in the continent, he notes their diversity in history, theology, and political orientation, and acknowledges the political ramifications of Pentecostalism in the continent.

A theme of particular interest in the book is fundamentalism. He argues that if viewed from "its classical Christian sense of denoting some belief in the Bible as inerrant, almost all African Christianity is fundamentalist." He highlights the key features of this movement in the US and contrasts the American setting with the African which makes common application of the term to both contexts unviable, and says, "it is not clear what application the description 'fundamentalist' has in sub-Saharan Africa, or what debate is advanced by employing the term." Consequently, he avoids the usage of the term. This is curious, especially against the backdrop of his earlier claims regarding African "fundamentalist" Christianity in *Christianity: To Save or Enslave?* and "Christian

Fundamentalism and Development," which are highlighted above. A possible explanation for this could be that by 1998, he had come to a more balanced evaluation of the movement (Burgess 2008, 233, footnote # 108).

In "The Complex Provenance of Some Elements of African Pentecostal Theology," Gifford examines some features of Pentecostal theology and relates the African manifestation of that theology to the global outlook. In doing so he uses three contemporary theological themes within Pentecostalism to advance his claim that the African version has two provenances: internal – "traditional preoccupations" – and external – North American influence. The three themes are faith gospel, deliverance, and Christian Zionism.

The last publication by Gifford to be highlighted is *Ghana's New Christianity: Pentecostalism in a Globalising African Economy*. The churches under review in this book, or what he calls "Ghana's new Christianity," are the Pentecostal ones. However, he chooses to describe them as "charismatic" because, according to him, "they are not the traditional Pentecostal denominations that have been in Ghana for up to seventy years." Rather they are the "newcomers." The churches, which are all located in Greater Accra, including Tema, Madina, and Adenta, are Mensa Otabil's International Central Gospel Church, Nicholas Duncan-Williams's Action Chapel International, Dag Heward-Mills' Lighthouse Chapel International, Nigerian David Oyedepo's Living Faith Church Worldwide, aka, Winners' Chapel, the World Miracle Church International of Charles Agyin Asare, and Elisha Salifu Amoako's Alive Chapel International. Gifford's avowed aims in the book are two: to analyse the identity and religious vision of this new Christianity and to participate in the debate on the socio-political role of the new Christianity. He posits that the main reason they flourish in Ghana is their "claim to have the answers to Ghanaians' existential problems and especially to their most pressing existential problem, economic survival."

He conveniently delimits the development of Ghana's "new Christianity" into four waves: the first began in 1979 with Duncan-Williams as the leading figure; the second is the teaching wave represented by Otabil; miracle healing characterises the third wave illustrated by Agyin Asare; while Salifu represents the fourth wave with its emphasis on prophecy. Nonetheless, he notes the difficulty inherent in carrying out a fine and distinct classification: "A complicating factor is the tendency for each succeeding wave to affect all existing churches, making 'pure' or 'non-hybrid' types hard to find."

According to him, the worship services of these churches have three parts: "praise and worship," offering, and sermon. The praise worship is usually led by a group of eight to fourteen musicians accompanied with

Western instruments such as electric guitars, electronic keyboards, trombones, trumpets and saxophones. Most of the songs are in English and are repetitive. The collection of offering is often accompanied with singing by the choir, made up mainly of young women in uniforms of blouses and skirts often varied on various Sundays. The sermon, or 'message', usually delivered extemporaneously by the preacher, takes the centre stage in their worship services and usually lasts for about an hour. Another feature of these churches highlighted by Gifford is their appropriation and dominance of the media, particularly TV and radio, and publication of Christian literature.

Having reviewed their growth and development, Gifford asserts, "by the end of the 1990s, Christianity was as dominant in Ghana as it had been twenty years earlier – indeed more so – but it was no longer the recognisably mainline (or 'orthodox') Christianity of old." Consequently, he says that these churches have had significant influence on Ghanaian Christianity such that there is the "charismatisation" of other churches, leading to the adaptation of some of their elements by the mainline churches in order to retain their members. Some of the mainline churches he mentions are the Methodists, the Catholics, the Presbyterians, and the Evangelical Presbyterian (EP), which even split partly over the issue. Its impact is also felt by the established Pentecostal churches. It has also attracted an element of defensiveness from the mainline churches as expressed in bumper stickers with such slogans as "I am a Methodist: I love my Church," "I love being an Anglican" and "I am a Catholic, I will remain a Catholic, and I will die a Catholic." Also, according to Gifford, the recurring emphases of these churches are success, wealth and status.

On the political role of Ghana's mainline churches, Gifford says that they are involved in education and development, the characteristic found in some Pentecostal denominations such as the Assemblies of God through their Relief and Development Services. In contrast, however, the newer churches spiritualise politics by way of "spiritual causality" which manifests in three ways. The first is an "enchanted approach" whereby demons, whose origin is seen in African Traditional Religion, are seen as being responsible for the political situation; hence their spiritual power must be broken. Biblical understanding, which has 2 Chron. 7:14 as key text and which believes that "the national plight is caused by apostasy, particularly failure to worship God properly on the part of the leader," is the second way. The third way by which Pentecostal churches spiritualise politics, according to Gifford, is the use of the faith gospel, or the "power of the word," to warn Christians against negative declarations on the condition of the nation. Gifford raises issues with all these strands as they downplay human agency in effecting socio-political change.

In all, the writings highlighted above cover a period of nearly fifteen years during which time Gifford maintained a consistent pattern in his valuation of African 'fundamentalist' Christianity regarding social witness. They are marked by some characteristic strengths. These include his method of data gathering, especially with its focus on primary sources and interaction with people at the grassroots level, especially their "songs and prayers, sermons and testimonies," which convey "their symbolic cosmos" (Gifford 1998, 27); the ethnographic description of some of the settings (e.g., see Gifford 1987, 63-92); and his analysis, which are quite engaging and detailed. In summary, they are "thoroughly researched, well documented, and rich in material" (Ter Haar 2006, 486) and possess "clarity of expression" (Kalu 2008, 198). Hence, his extensive study on African Pentecostalism has served as a trail blazer, thus opening the way for other works (Ellis 1998, 176).

Nevertheless, two preliminary responses are necessary here. First, in as much as it is true that there are traditional and occidental influences in the provenance of the phenomenon, Gifford leaves out a major factor in the reception of this form of Christianity among Africans, namely, their current *Sitz im Leben*. The present life situation of Africans is another major factor in the growth and spread of such churches as they are deemed to offer their members hope in the midst of despondency, faith in the context of general pessimism, and spiritual and emotional tools to compensate for the failures of the state (Ojo 2004, 2-3). However, this correspondence between the present African situation and the growth of the Pentecostal faith is not necessarily synonymous with causality as it is more complex than that. Ruth Marshall rightly notes this when she comments, "The relationship between crisis and religious revival is by no means a causal one, as many materialist or idealist approaches seem to suggest" (Marshall 2009, 8).

The second response is that in his historical overview of the movement, which he claims spread to the continent in the 1980s, Gifford fails to mention the earlier manifestation of the phenomenon in the early 1900s, such as the movement centred on Garrick Braide, aka, "Prophet Elijah II," in south-eastern Nigeria. According to Frieder Ludwig, Braide strictly kept to the Ten Commandments, engaged in rigorous lifestyle, extensive prayer and fasting, faith healing, and evangelistic preaching, and the movement he founded established schools (Ludwig 1993, 296-317; cf. Sanneh 1983, 180-184; Kalu 2008, 249). These are characteristics similar to the 'fundamentalist' brand of Christianity analysed by Gifford. Therefore he should have at least made reference to Braide and his movement. Further evaluation of Gifford's perspective on this brand of Christianity will be carried out in the last chapter.

Moreover, on the Western influence on African Pentecostalism, the following insight by Andrew Walls is instructive:

> The view is sometimes expressed that the recent growth of Pentecostal forms of Christianity in many parts of Africa is a product of a new North American missionary presence backed by massive North American money. The studies... do not support this view. North American Pentecostal mission in Africa had only a modest impact until the indigenous Pentecostal movements began. African Pentecostalism is very much aware of its world links, reads the North American literature and invites North American preachers, but it is financially self-supporting (Walls 1996, 146).

## 2.4 Summary of Gifford's View on African Evangelical and 'Fundamentalist' Christianity on Social Responsibility

Two pertinent questions that arise from the foregoing analysis are: What constitutes African evangelical and fundamentalist Christianity from Gifford's perspective? And what is their attitude towards social responsibility and development? Thus the aim of this section is to supply answers to these two questions from the publications reviewed above and thereby summarise his views on this form of Christianity. A number of these have already been highlighted in the previous section; yet, for the sake of emphasis and clarity, they are repeated hereunder.

His view of African evangelical churches could be gauged from *Christianity and Politics in Doe's Liberia*. While using the Liberian experience as a basis, he describes such churches as those that lay emphasis on allegiance to government, spiritual combat between good and evil spirits, and biblical fatalism that leads to an anti-political attitude towards the society. According to him, they also stress the depravity of the human nature, the evil of the world, and patience in affliction, with a focus on heavenly reward. In his view this form of Christianity "turned its back on any social awareness, insisted on a complete separation between Christianity and politics and purportedly left politics aside."

Therefore, to him, these qualities are not only political, in spite of their disdain for "politics," but inadvertently a "vote for the status quo." Thus, the main emphases of evangelical churches he identifies are unswerving loyalty to the government, spiritual warfare between good and evil spirits, biblical fatalism, human depravity, malevolent world, patience in affliction, and hope for heavenly reward.

He describes fundamentalism as "that sector of Christianity which in the US in the early part of this century reacted against developments in mainline churches, in particular against the historical study of the bible and against social involvement" (Gifford 1991, 10). Basically he argues that the "fundamentalist" form of Christianity emerged in America in the early 20th century and re-emerged as a significant trend in the late 1970s, and later spread to Africa in the 1980s. He examines the implications of this form of Christianity for the present crisis in the continent and weaves his argument around five major fundamentalist themes, namely, Dispensationalism, The Faith Gospel, Evil Spirits, The World, and The Human Person. He then contrasts its theology with ecumenical theology and alleges as follows: "And just as it is ecumenical theology itself that promotes involvement in development, we will argue here that it is fundamentalist theology itself that militates against development" (Gifford 1991, 10).

From a summation of his views on "fundamentalist" Christianity, as expressed in the publications under review, the characteristics of such Christianity he identifies include the following: emphasis on the faith gospel, dispensational millenarianism, Christian Zionism, belief in the imminent return of Christ, preoccupation with evangelism with focus on conversion of unbelievers, belief that the world is evil and therefore they must not be entangled with it, since "nothing this-worldly matters", belief in human depravity, and dualism of body and soul. He notes that this form of Christianity is gaining much ground in the continent. Thus, he asserts as follows: "Fundamentalist Christianity, while nowhere near numerically superior, is undoubtedly the area of greatest growth in African Christianity today" (Gifford 1991, 19).

A juxtaposition of the above characteristics of evangelicalism and fundamentalism by Gifford would show that evangelicals and "fundamentalists" share the same basic tenets from his perspective. This close affinity between fundamentalism and evangelicalism is underscored by H.V. Synan when he describes such affinity as "a movement among theologically conservative Protestant churches that reached its height in America during the 1920s and survives in resurgent post-WWII fundamentalist and evangelical churches and movements." Continuing, he says that fundamentalism is known as evangelicalism in its more respectful form (Synan 2002, 655). Anthony Balcomb also draws the same conclusion about this form of Christianity when he broadly defines evangelicalism as "that kind of Christianity, emerging from the Pietist stream of the Reformed tradition, whose emphasis is on 'salvation through personal encounter with the risen Christ.'" He includes Pentecostal and Charismatic movements and other churches that insist on personal salvation and Christian discipleship through adherence to Scripture within this group (Balcomb 2004, 146-147). David Maxwell is on a similar template when he describes African "born-again" Christianity (his term for African "fundamentalism") as comprising of "a diverse coalition of evangelicals, charismatics and Pentecostals, who share a belief in the infallibility of Scripture and, more importantly, stress the centrality of a 'born-again' experience." Continuing, he adds "a third defining characteristic," namely, possession of the 'gifts of the Spirit demonstrated through healing, *glossolalia*, exorcism and prophecy" (Maxwell 2002, 160).

The three churches in this study all share these attributes as they hold to the four tenets of evangelicalism: the authority and sufficiency of the Scriptures, the uniqueness of redemption through the death of Christ upon the cross, the need for personal conversion, and the necessity, propriety and urgency of evangelism (cf. McGrath 1997, 121-122). Also, there is evidence of belief in the "five points of fundamentalism" among them, namely, verbal inerrancy of the Bible, the divinity and virgin birth of Christ, substitutionary atonement, the physical resurrection of Christ, and his corporeal and personal return (Lechner 1998, 197). The link between Pentecostalism in Africa and "fundamentalism" is also noted by Max Assimeng who claims that "Pentecostalism is the most recent large-scale manifestation of Christian sectarianism to reassert the near-orthodox fundamentalism position of Arminian Protestantism" (Assimen 1986, 133).

However, this form of Christianity receives scathing criticisms from Gifford regarding its attitude towards social responsibility and development because of its alleged preoccupation with winning souls. He says, "according to this Christianity, nothing matters but evangelisation" since "a Christian's task is to prepare others for the imminent return of Jesus" (Gifford 1991, 12; cf. Peter 1994, 104-105); hence there is "lack of social responsibility in this prosperity gospel" since "only Christians matter" (Gifford 1991, 14; cf. Gifford 1990, 9).

Going further, he alleges that not only is this form of Christianity withdrawn from development but that it attacks the Christians who do: "So fundamentalist Christianity not only withdraws its adherents from involvement in development; it attacks Christians who are so involved" (Gifford 1991, 13). Hence, he asserts that from a fundamentalist perspective, "the plight of the nation, or the state of society, or the situation of the general population (is) of no importance to the Christian" (Gifford 1991, 14); and adds,

> These Christians have almost no understanding of human responsibility in relation to surrounding circumstances. There is no emphasis on taking

> charge of one's own life, using one's natural faculties, energies and g(i)fts to collaborate with others in projects calculated to control one's own life and world. Such an attitude can be positively rejected (Gifford 1991, 17).

His prognosis of Africa's development from a fundamentalist perspective is negative; hence, he boldly declares: "the fundamentalist Christianity spreading in Africa today is not a spur to Africa's development. On the contrary, it leads its adherents to downplay the importance of development, to dismiss it as irrelevant, or even positively to turn their backs on it" (Gifford 1991, 17). It is noteworthy that Sylvester Odion-Akhaine makes a similar claim about Pentecostal churches in his comparative evaluation of Pentecostal and Catholic approach with regards to speaking the truth to the political powers in Nigeria when he says, "In our humble opinion, the Pentecostalists are less forthcoming and more obsessed with miracle-induced prosperity than interrogation of the structures of society that create condition of poverty" (Odion-Akhaine, 2006, 759-760).

What is the reason for the above posture by fundamentalist Christians? Gifford attributes it to their theology and worldview, particularly their eschatology, cosmology, anthropology, and missiology in which they are ultra-fixated with other-worldliness. His interpretation of three of these themes will suffice. The first is their eschatology. According to him, due to their millenarian dispensationalist orientation, these churches place so much emphasis on the end-time return of Christ that it leads them to three tendencies: (1) a passive acceptance of social ills as part of the outworking of God's will and as precursors to the end-time since "nothing else matters but preaching the gospel, or fulfilling the Great Commission (Mat 28, 19)" (Gifford 1990, 4); (2) anti-socialism; and (3) Christian Zionism, which accords Israel 'blind' support.

The second theological basis of the "fundamentalist" posture towards the society he identifies is their cosmology. According to him, their belief is that the world is evil. He explains this understanding as follows:

> Since the world is opposed to God, Christians have no other duty than to turn their back on it, to flee it, to keep themselves unspotted by it. The church, made up of born-again believers, becomes an alternative society; the church is the place to control, to order, and to call home. The church is where God is to be found, where God acts. The church is God's domain; the world is the domain of Satan (Gifford 1991, 15-16).

Hence, he says, "nothing this-worldly matters" to them (Gifford 1991, 16).

The third example is their anthropology, which Gifford claims is dualistic: "Fundamentalist Christianity is dualist, and one of the most deep-seated dualisms is that between the body and the soul." They also teach "the total depravity of the human person." Therefore, he concludes,

> These Christians have almost no understanding of human responsibility in relation to surrounding circumstances. There is no emphasis on taking charge of one's own life, using one's natural faculties, energies and g(i)fts to collaborate with others in projects calculated to control one's own life and world. Such an attitude can be positively rejected (Gifford 1991, 17).

Rather, he claims, as a consequence of its focus on supernatural causes and personal healing and holiness, this strand of Christianity diverts attention from the economic and political factors responsible for social ills and therefore distracts from social involvement (Gifford 2005, 86).

Gifford notes that this form of Christianity is making inroads into the mainline churches mainly through popular literature and, then, alleges, "Popular theology which stresses holistic salvation, human dignity, human potentiality, social responsibility, structural awareness and development is hard to find in Africa" (Gifford 1991, 18). Nevertheless, he notes that some of these churches indeed engage in development projects, but adds that they are few and that doing so is "really a contradiction of their teaching, for their message itself militates against such commitment" (Gifford 1991, 19).

However, it is pertinent to note that there are other scholars of African Christianity with nuanced positive assessment of its Pentecostal and 'born-again' strand. Two examples will suffice. The first is Richard Burgess. In his study on the nature of Nigerian Pentecostal theology vis-à-vis its contributions to intercultural theology, Burgess rejects Gifford's argument that the new African Pentecostal churches dismiss Christian tradition, demonise traditional African religion and culture, and reject the current socio-political situation as theologically irrelevant. Instead, he sees it as an important source for African theology and argues,

> African Pentecostalism appeals to popular religious sensibilities precisely *because* it resonates with the pragmatic and power-oriented nature of African indigenous spirituality, while at the same time allowing individuals to break free from the religious and social ties of the past and construct new identities for themselves. By doing so, it helps to alleviate the dilemma of dual allegiance to church and traditional cult so prevalent in mainstream African Christianity (Burgess 2008, 30-31).

The second nuanced assessment of African Pentecostalism, especially on its political role, to be mentioned is from Ruth Marshall. She sees Pentecostalism as "the most important socio-cultural force in southern

Nigeria." Using the concepts of collective "political redemption" and "political spirituality," or the "will to found anew," she argues that the Pentecostal movement "has as its principal aim a project of individual and collective renewal and regeneration through a process of conversion based on the idiom of new birth." Continuing, she adds, "The project of conversion involves the elaboration of new modes of government of the self and of others, in which practices of faith are fostered by specific disciplines of the body and the mind, emphasizing purity, rectitude, righteousness, and interiority" (Marshall 2009, 2-3). Such an understanding of conversion is not bereft of political agenda for it

> expresses the political ambition of replacing a corrupt regime with a new form of righteous authority that presents itself as the unique path to individual and collective salvation… conversion is represented as a means of creating the ideal citizen, one who will provide a living incarnation of the *nomos* of a pacified and ordered political realm (Marshall 2009, 14).

Thus, in contrast to Gifford's claim that such Christianity is apolitical, Marshall shows that it is inherently political in its theology by virtue of its concept of conversion. The full import of this will be appreciated in the discussion in Chapter Six on the theology of conversion of the three congregations in this study.

## 2.5 Gifford in Select Literature on African Christianity

The aim of this section is to explore how Gifford is perceived among scholars of African Christianity. A close examination of a number of their works would yield at least three categories. However, due to space, only a few of them will be mentioned.

The first group has to do with those who merely reference his works in the bibliography without citing or interacting with him. Daniel Smith is an example of this (cf. Smith 2001). However, it is curious that in a work written in 2001 and which explores the "issues of prosperity, power, and inequality in the realms of Pentecostalism, the occult, and the patrimonial system of patronage that dominates Nigeria's political economy," no mention is made in the discussion to at least Gifford's *African Christianity: Its Public Role*. This observation is based on Gifford's pedigree on the topic as earlier highlighted.

The second set, represented by Matthews Ojo, Rosalind Hackett, Marleen De Witte, and Birgit Meyer, are those who merely cite and use him as a basis to substantiate and/or expand their arguments. For example, while Ojo uses his work to elaborate on his review of the role of religion in Africa's politics and governance (Ojo 2004, 1), Hackett cites him in her defence of the variegated nature of Pentecostal interpretation of 'prosperity', as well as her review of Bonnke's "Cape to Cairo" crusades in Africa (Hackett 1998). The same could be said of De Witte in her exploration of "the sonic sacralization of urban space in the multicultural city of Accra," wherein she merely cites him to note the new form of Pentecostalism that sprung up in Ghana (De Witte 2008). Meyer accepts Gifford's categorisation of the respective responses of the mainline churches in Ghana and the Pentecostal ones to the law requiring all religious bodies to register with the National Commission on Culture, as well as their particular attitudes towards politics in the country (Meyer 1998).

The third category of African scholarship on Gifford is those who, though they endorse him at certain points, yet engage with and/or dispute his methodology and/or claims. Five examples stand out, namely, Jo Sullivan, David Maxwell, Allan Anderson, Kwame Bediako, and Ogbu Kalu. In his insightful and concise review of Gifford's *Christianity and Politics in Doe's Liberia*, Sullivan notes its "several errors of fact," as well as its use of "inappropriate tone" (Sullivan 1995, 368). However, Sullivan only mentions Gifford's perspective on evangelical churches in Liberia without engaging with it. Maxwell, while on the one hand endorses Gifford's position on the negative influence of American Pentecostal preachers on southern Africa due to their impact in "diverting attention away from the socio-structural causes of the region's ills" (Maxwell 1998, 350), he paradoxically, on the other, disputes his dismissal of Pentecostal ability to aid Zimbabweans in grappling with the pressures of modernisation. Thus, he argues, "Gifford's dismissal of pentecostalism in favour of Latin American style liberation thereby misses the point that Pentecostalism (cf. Gifford 1987) does enable ordinary Zimbabweans to face painful social and economic transitions" (Maxwell 1998, 369). Furthermore, in his review of another of Gifford's books, *African Christianity: Its Public Role*, Maxwell accepts its richness and originality but critiques some of his conclusions as being "less than convincing," since they "arise from an inadequate methodology and an inattentiveness to the sheer complexity of the culture and the history of African religion" (Maxwell 2000, 478). Also, Anderson accuses him of ignoring "some fundamental features of Pentecostalism, now predominantly a Third World phenomenon, where experience and practice are more important than formal ideology or even theology" (Anderson 2007). Bediako, while commenting on *African Christianity*, has this to say: "For all its rich documentation and judicious commentary in a number of places, Gifford's study is disappointing, precisely because of ... a failure in perspective" (Bediako 2005, 118).

On his part, Kalu uses Gifford's series of studies on Ghana's "New Christianity" to illustrate "the instrumentalist discourse" on African Pentecostalism, a discourse he says "explores how religion serves as an instrument to achieve other goals that provides political, economic, and psychological adjustments to new realities by vested interests" (Kalu 2008, 187). He credits Gifford for his familiarity with the Ghanaian territory, his capacity for in-depth research, and his interaction with the growing literature (Kalu 2008, 192). He also raises issues with his methodology, which he argues portrays a skewed image of some Pentecostal leaders, for example, Dag Heward-Mills and Abu Bako, and "devotes great uncritical attention to cataloguing the discomfort of secularists, Muslims, journalists, and synods of mainline churches about the so-called false prophets." Thus, Kalu metaphorically describes it as "a thick description of dancers without hearing their music," as a result of Gifford's outsider perspective that fails to seriously consider the insider viewpoint (Kalu 2008, 198-199).

From the above analysis, it is apparent that none of the studies deliberately set out to empirically test Gifford's claim that African evangelical and "fundamentalist" churches are anti-development and lack social responsibility with a view to either confirming or rebutting it. This is what this book seeks to do, using data collected from field research in three local congregations in the Niger Delta region of Nigeria on their witness and theology within the context of crude oil and conflict. Specifically, the claims to be ascertained are the following:

1. That such churches lack social responsibility, since "the plight of the nation, or the state of society, or the situation of the general population (is) of no importance to the Christian," and are therefore apolitical;
2. That they encourage a passive acceptance of disasters and misfortunes;
3. That they lack a commitment to development and that, even when they engage in development projects, it is a contradiction of their teaching;
4. That they place emphasis on unswerving allegiance to government, the evil of the world, spiritual combat between good and evil spirits, dualism of body and soul, and heavenly reward, since "nothing this-worldly matters" to them;
5. And that they are preoccupied with evangelism, or soul winning.

# Chapter Three

# Political Economy of Oil and Conflict in Nigeria

## 3.1 Introduction

As will be shown in this chapter, the conflict in the Niger Delta is a multi-faceted phenomenon with a number of trajectories. These trajectories represent the various perspectives and interest groups in the crisis. Thus, the chapter will outline Nigeria's political economy of oil and conflict by describing the key parties and how they interact, as well as their respective economic perspectives to crude oil. It will also critically evaluate the perspectives. The objective is to give the socio-political background to and facilitate the achievement of the second main goal of this book – the assessment of the validity, or otherwise, of Gifford's claims on African evangelical Christianity regarding social responsibility.

## 3.2 The Political Economy of Oil and Conflict in Nigeria

Nigeria's political economy of oil and conflict could be delineated into a number of perspectives within a grid of assessment. The grid ranges from pro-industry viewpoints and practices to pro-local people perspectives and activities. From personal experience and close observation of the crisis spanning a period of more than fifteen years and an in-depth study of the violence from primary and secondary sources, the grid has the Federal Government of Nigeria and the multi-national oil companies, or MNOCs, on the pro-industry end, and civil society and youth groups at the pro-local people end. In-between the two ends are political and community leaders in the region. These are described below.

### *3.2.1 The Federal Government and the MNOCs*

The first perspective in Nigeria's political economy of oil to be mentioned is the pro-industry perspective. Its main goal is the maintenance of the status quo in Nigeria's oil economy and has as its primary interest the continued exploration, extraction and exploitation of the oil resources in the region. It is represented mainly by the Federal Government of Nigeria and its agencies, such as the Nigerian National Petroleum Corporation (NNPC) and the Niger Delta Development Commission (NDDC), and the MNOCs. The NNPC was established in 1977 following the merger of the defunct Nigerian National Oil Corporation (NNOC), which was created in 1971, and the Federal Ministry of Petroleum and Energy. According to T.E. Turner, this merger followed the approval of the recommendation to the Federal Executive Council by the Ministry of Petroleum and Energy following years of wrangling between the two agencies of government (Turner 1977, 209-210). It has eleven subsidiaries and manages Nigerian government's equity shares in the Joint Venture Partnerships. The NDDC was established via an act of the National Assembly in July 2000 as an interventionist agency of state for the development of the oil endowed areas. It has antecedents in the Niger Delta Development Board (NDDB) of 1961 and the Oil Mineral Producing Areas Development Commission (OMPADEC) of 1992.

The NDDB was belatedly formed in 1961, following the recommendations of the Henry Willink's Commission in 1958 with the mandate to develop the region. The Commission was set up by the Secretary of State for the Colonies to look into the fears of the ethnic minorities in Nigeria and recommend ways of handling them. Its report is entitled *Report of the Commission Appointed to Enquire into the Fears of Minorities and the Means of Allaying Them* and was presented to the Parliament by Alan Lennox-Boyd, then Secretary of State for the Colonies by Command of Her Majesty in July 1958. It was chaired by Henry Willink, hence its designation as "Willink's Commission." But the Board could not do much in its seven years of existence. Thus, in 1992, the Babangida-led military government formed the OMPADEC. But due to "lack of focus, inadequate and irregular funding, official profligacy, corruption, excessive political interference, lack of transparency and accountability, and high overhead expenditure" (Federal Republic of Nigeria, 102), the Commission was dissolved in 1999 and replaced with the NDDC in 2000. The NDDC is primarily mandated to "formulate policies and guidelines for the development of the Niger-Delta area" (Federal Republic of Nigeria 2000, 7). As will be shown below, it has already received negative appraisals from certain quarters. At least seventeen foreign oil companies operate in Nigeria, with five predominating, namely, SPDC, ExxonMobil, ChevronTexaco, TotalfinaElf, and Agip. These are complemented by the activities of thirty-eight indigenous companies, or private operators (Khan 1994, 35).

Jedrzej Frynas has demonstrated that the Nigerian Government and the MNOCs, for mutually beneficial reasons, work together as partners (Frynas 1999, 8), or as "the coalition of oppressors in the axis of greed" (Kemedi 2003, 134-135). For example, whereas the companies provide the government with revenue, the latter provides them with access to crude oil by way of licences and the enabling environment and framework within which to operate (Frynas 1999, 8), as well as protection from attacks by local communities (Zalik 2004, 404). As partners they use various means and tactics in their quest to maintain the status quo. Some of the methods used by the Government and its agencies include: allocation of certain percentage of oil revenue to the oil producing states, which has ranged from 1.5% in 1982 to 3% in 1992 and 13% since 1999; coercion and use of force, especially through the joint-security outfit, the Joint Task Force (JTF), code-named 'Operation Restore Hope'; exploitation of the ethnic factionalism in the region; and "stakeholders" meetings. The politics of "stakeholders" meeting was one of the major tactics adopted by the Obasanjo administration, especially between 2003 and 2004, in its attempt to douse the tension in the region.

On their part, the MNOCs adopt various means to appease the people of the region, especially the youths, so as to enable them continue with their production activities. Some of the measures include corporate social investment, such as community development projects and the building of the "social and human capital and capacity of the people" (Total E&P Nigeria 2009, 7), cash payments to community leaders and youth groups, and "ghost worker" salaries for youths. "Ghost workers" refers to non-existent staff represented by fictitious names included on the payroll of some companies and/or Government ministries, parastatals and agencies (Brooks 2004, 27).

In dealing with the specific issue of threat to their operations, the companies adopt certain strategies. Kenneth Omeje describes the strategies as security "communitization," security privatization, and security corporatization (Omeje 2006, 486-488). Security communitization is "the contractual engagement of members and youth groups of the local oil communities to provide security for oil installations and operations within local communities"; "security privatization" refers to "the surge of specialized security companies/organizations and private military corporations (PMCs)"; while "security corporatization," or "corporate militarism," is "the increasing empowerment of some large business corporations to operate their own security outfits or to considerably run (equip, fund and/or command) a detachment of the state's defence forces assigned to protect the corporation's personnel and property" (Omeje 2006, 486-488).

In addition, the MNOCs utilise the "divide-and-rule" tactic, or manipulation, in their dealings with the local communities that are embroiled in communal strife as they "have often covertly and more openly supported one or both of the fighting parties." According to Dimieari Kemedi, the aim of the tactic is to deflect attention from them and ensure "that the people remain divided and concentrate on mutual annihilation rather than question in any effective and efficient manner, the unjust system of exploitation," which forms the basis of their activities (Kemedi 2003, 134).

In terms of attitude or economic paradigm, the major emphases of this perspective are *resource ownership*, *resource allocation*, *resource management* and *resource exploitation*. *Resource ownership* refers to the bestowment of the ownership and control of all land in Nigeria and all petroleum resources within its borders and its territorial waters with the state. This is enshrined in a number of Nigerian laws, such as the Oil Pipelines Act 1956, Oil in Navigable Water Ways Act 1968, and especially the Petroleum Act 1969 and the Land Use Act 1978.

The second attitude towards crude oil from this perspective is *resource allocation*, or "revenue allocation," based on the principle of derivation. It has been a highly controversial issue in Nigeria's history, especially since the late 1960s when the country turned from an agrarian economy to oil based one. The principle has had a chequered history in its evolution: in 1953 it was one hundred per cent, fifty per cent in 1960, and forty five per cent in 1970. It took a downward trend in 1975 to twenty per cent, two per cent in 1982, and then reached its lowest point by 1984 when it was 1.5 per cent. It was slightly increased to three per cent in 1992 and was further increased to thirteen per cent in 1999.

*Resource management* is the third paradigm of the conservative perspective. As a result of the discontent and agitation expressed by the people of the region regarding the inequitable share of oil revenue allocated to them, coupled with the high level of unemployment, poverty and poor infrastructure, the government adopted another route to meet the unique developmental needs and challenges of the region. This is in the form of special remediation and interventionist agencies entrusted with the mandate to meet those needs. The successive agencies as earlier highlighted are the NDDB of 1961, the OMPADEC of 1992, and the NDDC of 2000.

The last paradigm to be mentioned is *resource exploitation*. This is represented by the NNPC, the MNOCs, and local companies. As agents of international capital, these companies, especially the MNOCs, focus primarily on maximising profit and making handsome economic returns for their share holders. Thus, their emphasis is on resource exploration and exploitation. They have also been accused of being agents of international capital and of colluding with the Nigerian state in marginalising the people of the region and degrading the environment (Onoh 1983, 88; Ikein 1990, 87; cf. Northcott 2007, 95-96; and Litvin 2003).

### *3.2.2 Political and Community Leaders in the Niger Delta*

The second set of people in the grid of Nigeria's political economy of oil and conflict is the political and community leaders in the region. This is seen in most of the Governments of the States and local chiefs in the region. They canvass for the modification of existing structures without fundamental changes. Their major call is for substantial increase in the derivation fund from the present 13% to at least 25%. This is illustrated by the demand for an increase to 25% in the derivation fund made by the south-south political leaders, and supported by the south-east and south-west leaders, at the National Political Reforms Conference (or National Dialogue) formed by President Obasanjo on February 21, 2005. The North opposed the demand and this led to a stall in the plenary session of the Conference (Ajayi 2006, 126).

Yet this perspective has not given proper account of the management of the thirteen per cent oil derivation revenue allocated to the oil producing states since 1999, which led to a substantial increase in the volume of resources that was available to such states. Moreover, several local chiefs are accused of high-handedness and misappropriating community development funds meant for their communities, leading to loss of faith in them by the youths. This accusation was expressed by a large number of individuals interviewed during field research, including "ordinary" men and women, academics, chiefs, pastors and oil company workers. Also, in a speech to handover the report of his Commission to the Rivers State Government on March 10, 2009, the Chairman of the Truth and Reconciliation Commission set up by the State Government in November 2007, Hon. Justice Kayode Eso, has this to say about some local chiefs:

> We… discovered that some chiefs turned themselves into tyrants. They will not listen to the complaints brought against them or against those in their good books in the community… These chiefs believe they could rule and indeed did rule with impunity. They refer to themselves as Majesties

and Highnesses, which of course they are, but they have gone beyond this (Eso 2009, 41).

Moreover, between November 2004 and October 2005, the cumulative revenue from the Federation Account that accrued the nine oil endowed states from the statutory allocation and the thirteen per cent derivation fund according to the Federal Ministry of Finance was about ₦521.8 billion, or US$4.1 billion. Out of this amount ₦291.5 billion, or US$2.3 billion, was from the derivation fund in which four of the states with the largest oil deposits, namely, Rivers, Bayelsa, Delta and Akwa Ibom, received a total of ₦263.3 billion, or US$2 billion, between them. When evaluated from the bigger picture of the thirty six States and Abuja, what these States received during the period, especially Rivers, Bayelsa, Delta, and Akwa Ibom, dwarfs what the non-oil producing States received.

The major economic paradigms of this group regarding crude oil are mainly *resource allocation* and *resource management*, with a modicum of emphasis on *resource control*. Already resource allocation and resource management have been described, only to add that at the state level, they refer to the receipt and management of the states' share of the federally collected revenue and the thirteen per cent derivation that accrue the States and Local Government Areas in the region. This group vacillates between pro-industry and pro-local people, depending on its economic and/or political agenda.

The paradigm of *resource control* refers to the demand of the people of the Niger Delta to have control over the management, exploration and exploitation of mineral resources in their region (Wokoma n.d., 182-183). Itse Sagay has outlined the three major components of the idea: the power and right of a community or state to raise funds by way of tax on persons, matters, services and materials within its territory; the exclusive right to the ownership and control of resources, both natural and created within its territory; and the right to customs duties on goods destined for its territory and excise duties on goods manufactured in its territory (cf. Sagay 2007). The demand grew as a result of several factors: the failure of State to equitably distribute the revenue from oil to the region from where it is produced, years of neglect and marginalisation meted to the region, and the environmental degradation occasioned by oil exploration.

### *3.2.3 Civil Society Groups*

In his insightful analysis of civil society, Augustine Ikelegbe has indicated that civil society is the main analytic paradigm in African politics today as it is "romantically associated with the wave of popular protests and social mobilisations that has resulted in democratisation since the early 1990s." He defines it as "the composite of organisations of citizens, in usually non-partisan and non-profit associations, that is situated and intermediating between the family and society on the one hand and the government and state on the other"; or as "the aggregate of institutions whose members are engaged in a complex of non-state activities." Consequently it is characterised by distinctness and independence from the state, with objectives that are basically civil and public and activities that are marked by claims, contestations and popular struggles (Ikelegbe 2001, 2; cf. Bayart 1993, 109-125; World Bank 1989; Hyden 1992).

Ikelegbe's comments above are relevant to the Niger Delta question as a number of civil society groups exist in the region with support for the local people on the question of oil. Some examples of the civil society groups in the region include: the CHIKOKO Movement, led by Oronto Douglas; the Council of Ijaw Associations Abroad; Ijaw National Congress (INC); Concerned Ijaw Youths for Development of Bayelsa and the Niger Delta; Bayelsa Indigenes Association; Niger Delta Women for Justice; Isoko Youth Movement; the Niger Delta Congress; the Movement for the Survival of Ogoni People (MOSOP); and Ugborodo Women of Escravos (Ikelegbe 2001, 12-14; cf. Okafor 1999; Ikwunze 1999, 1-2). These groups generally advocate self-determination, fiscal federalism, equity, and resource control and adopt variegated methods in their advocacy, ranging from violence to active non-violenct resistance. Some of these civil society groups normally follow non-violent means in their struggle against the perceived oil injustice, while others follow violent means. The use of violence mainly characterises the youth groups, which are taken up further in the next subsection. Two normative non-violent groups within this perspective are MOSOP and the Ugborodo women of Escravos in Delta State. Consequently, they deserve further comments.

MOSOP is a non-governmental organization of the Ogoni people formed in 1990. Its main objective is the securing and protection of the economic, social, cultural, and environmental rights of the Ogoni people, mainly through non-violent means. It also advocates similar rights for other minority groups in Nigeria (Pyagbara 2006). Ledum Mitee, its one-time President, explains that the non-violent approach was deliberately chosen by the group after a series of meetings wherein it considered all possible options, including sabotage of oil facilities and guerrilla warfare. According to him, the factors that informed their choice of non-violence include: the Christian teaching on non-violent resistance; their geographical terrain that made guerrilla tactics unviable; their small population in relation to the rest of the country; the conviction that using the same tactics and weapons of one's enemy, such as violence, was wrong; their desire to draw international sympathy and support to their cause; and pragmatism (Mitee 2007).

The role of women's groups in the region against the backdrop of the perceived injustice in Nigeria's oil economy is encapsulated by the Ugborodo Women's Protest of 2002. Ugborodo is one of the communities in Escravos, Warri South Local Government Area of Delta State. Iyenemi Wokoma gives a helpful explanation of this protest and its antecedents: Decades of oil exploration and exploitation by Chevron had allegedly led to environmental degradation and pollution of the area, which destroyed the people's means of livelihood, namely, fishing and subsistence farming. This resulted in a high level of poverty and resentment from the local community. The resentment reached its peak in June 2002 when a group of seven hundred women met and subsequently wrote a letter to Chevron requesting for a clean environment and job opportunities for their youths. A month passed without any response from the company. Then, on July 8 2002, "on their own and not accompanied by men, they invaded the Chevron platform at Escravos and brought operations at the facility to a standstill, promising to leave only after they have spoken to the Managing Director of Chevron" (Wokoma n.d., 176). The siege at the facility, which lasted for ten days and led to a production cut of 450,000 bpd throughout its duration, initiated a month-long protest against Chevron and Shell by 2000 women. The weapons they used in the course of the protest were food, music and dance. Eventually their non-violent active resistance succeeded in convincing Chevron to send their senior executives for negotiation. The dialogue led to an agreement between the two parties in which the company acceded to "employing more local people, funding schools, electricity and other infrastructure projects, and assisting the women in setting up poultry and fish farms" (Wokoma n.d., 177-178).

In summary, the non-youth Niger Delta civil society perspective in Nigeria's political economy of oil and conflict is characterised by its grassroots orientation, its non-violent ethos, and its calls for self-determination, resource control and fiscal federalism, environmental responsibility, and employment opportunities for the people of the region. Its dominant economic paradigms are *resource ownership* and *resource control.*

### *3.2.4 Youth Groups*

The last perspective to be mentioned is that of the various youth groups in the region. They mainly insist on a radical change in the existing political and economic structures in Nigeria. Some of the groups include the Ijaw

Youth Council (IYC), an umbrella body of Ijaw youth associations in Rivers, Bayelsa and Delta States formed in 1998 and which issued the Kaiama Declaration of December 11, 1998; the Movement for the Emancipation of the Niger Delta (MEND); the Niger Delta People's Volunteer Force (NDPVF); the Martyrs Brigade (MB); and the Coalition of Militant Action in the Niger Delta (COMAND). Their membership is drawn mainly from youths who feel frustrated by lingering unemployment and the perceived indifference of the Nigerian Government and the oil companies to their plight. Their main mode of operation is violence carried out through hostage-taking, sabotage and destruction of oil facilities, such as oil platforms and pipelines, violent attacks on agencies of state, such as armed forces personnel and military barracks, and inter-gang warfare. Hence these groups have variously been tagged "area boys," "vigilantes," "armed robbers," "cultists," "armed gangs," "hoodlums," "rebels," "militias," and "criminals" (Ukeje 2004, 605-617; cf. Osuigwe 2007), while they see themselves as freedom fighters, or liberators. The diversity in the composition of these groups further accentuates their complexity and calls for prudence in any attempt to classify and/or respond to them. This issue is further taken up below under the dimensions of the conflict.

The major demands of this perspective include convocation of a Sovereign National Conference, resource control, or fiscal federalism, and the abrogation and/or amendment of certain laws that regulate the oil industry in Nigeria. In its extreme form, some individuals who espouse this perspective have called for a break-up of Nigeria, which they describe as "the satanic and evil Nigerian state," into the pre-colonial sovereign states composed of individual ethnic groups, and this was presaged by Isaac Adaka Boro's 'Twelve Day Revolution' of 1966. Since they are many and represent a number of interests, they often fight among themselves, especially for political supremacy and/or control of the "illegal oil bunkering" routes in the region. According to Bronwen Manby, oil bunkering by the youth groups is carried out with the connivance and support of senior state and federal government officials (Manby 2004, 2).

However, as Kemedi has indicated, it is pertinent to note that in the history of the struggles of the people of the region against external oppression, violence has not been their main option, apart from the Akassa War of 1895, which the Nembe people fought against the British, and Isaac Boro's Twelve-Day Revolution of January 1966. In contemporary times they have generally not resorted to violence in dealing with their oppressors (Kemedi 2003, 136). According to Mitee, the resort to violence in the present struggle by the youth groups in the region took years to develop and is based on a feeling of frustration and the notion that "violence is the only language government understands." This notion grew out of the belief that non-violent forms of the struggle for justice and equity are often ignored by the government and the MNOCs; hence, violence is considered to be the next logical option (Manby 2004, 2). Mitee explains this feeling of frustration as follows:

> It's only when people are taken hostage, or someone is killed, or they've blown up some oil facility that the government wants to do something, or seems to be doing something in response. That has been counter-productive because many people now feel that the best way to get attention is to go violent… our own youths are feeling that they're being ignored because they are not violent or we're asking them not to be violent (Mitee 2006; cf. Numbere 2006).

Consequently, the introduction and intensification of the violence by the various youth groups in the past decade, especially, has achieved certain results for the region.

First, it has shown that despite their so-called minority status, the people of the region are capable of pressing for their rights, contrary to the insinuation that they are too small in number to threaten Nigeria's stability or economic wellbeing. Second, it has helped to heighten awareness of the problems associated with the oil industry in the region. Third, it has shown that the Nigerian state cannot afford to persist in ignoring the Niger Delta question. Fourth, it has demonstrated to the MNOCs that they can ill-afford to continue to take the people of the region for granted. Finally, it has led to a major policy change by the Federal Government from aloofness and arrogance to engagement with civil society groups and the people of the region (Osuigwe 2008, 11).

The last point above is given further credence by two measures taken by the Federal Government, namely, the formation of a 44-member Technical Committee on the Niger Delta in September 2008 and the unilateral granting of amnesty to militants in June to October 2009. The Committee, which was chaired by Ledum Mitee, was mandated, among other things, "To collate, review and distil the various reports, suggestions and recommendations on the Niger Delta from the Willinks Commission Report (1958) to the present, and give a summary of the recommendations necessary for government action" (Technical Committee on the Niger Delta 2008, v). The unconditional amnesty package was proclaimed by the late President Yar'Adua on June 25, 2009 and was to last from June 25 to October 4, 2009. According to the presidential declaration, the programme was for "all persons who have directly or indirectly participated in the commission of offences associated with militant activities in the Niger Delta," including those under detention and those already being tried for various offences in the region. This is predicated on their "willingness and readiness" to surrender their weapons, renounce violence "in all its ramifications unconditionally, and depose to an undertaking to this effect" (Ya'Adua 2009). The amnesty programme apparently led to the surrender of "more than 80 percent of rebels who have disrupted oil operations in Nigeria's Niger Delta" (Ekeinde 2009), including some of the key leaders, and they include Henry Okah, Victor Ben Ebikabowei, alias Boyloaf, Government Ekpemupolo, alias Tompolo, Prince Wisdom Amachree, the second in command to Mujahideen Dokubo-Asari of the Niger Delta People's Volunteer Force (NDPVF), Charles Kuboro, Nwaeze Adele, Kile Selky Torughedi, alias "Young Shall Grow," Farah Dagogo, Soboma Jackrich, and Christian Don Pero. However, due to its complexity, the youth perspective requires further in-depth analysis so as to bring out its dimensions.

## 3.3 Dimensions of Youth Militancy in the Niger Delta

The militancy in the Niger Delta is an apt illustration of an oil-based conflict by virtue of its ingredients and dimensions. Michael Ross has identified certain ingredients of such a conflict, such as large-scale availability of crude oil, large number of minority ethnic groups with inadequate compensation and disproportionate share of the cost of production, and wide geographical spread of oil facilities. Continuing, he argues that matters are made worse in the case of the Niger Delta by the high rate of poverty, unemployment, swampy terrain of the area, and the challenge of traditional authorities by youth groups (Ross 2003, 12-13).

Another factor that is germane to the militancy, but which is omitted by Ross, is the widespread availability of small arms and other lethal weapons (Human Rights Watch 2005, 2). The weapons come from various sources, such as rogue elements in the Nigerian military and international syndicates on the high seas (Best and Kemedi 2005, 12-44; Ginifer and Ismail 2005, 2; and Institute for Peace and Conflict Resolution 2002, 13). Consequently, the various dimensions of the conflict, which have already been anticipated in the preceding discussion on the various perspectives, are the political, the cult-related, the economic, the ethnic, the communal, and the criminal. The variegated nature of the dimensions of the militancy in the Niger Delta indicates the diversity of its motivations, such as agitation for recognition, struggle for justice, rebellion against the State and against autocratic, corrupt and gerontocratic traditional leadership, and greed. The dimensions are analysed hereunder.

### *3.3.1 The Political Dimension*

This is the aspect of the militancy in which politicians exploit the general situation of poverty and widespread youth unemployment to organise, arm and sponsor youth groups in the region in their quest for power. This was most prevalent in the run-up to the 1999 and, especially, 2003 general elections and subsequent attempts by the so-called winners to consolidate power by intimidating and terrorising their opponents. Their major mode of operation is the formation of neighbourhood or state-wide vigilante groups ostensibly for warding off criminals. As the Institute for Peace and Conflict Resolution, Abuja states, "These unconventional outfits are used for political thuggery, destruction and vandalisation of public property, and assaults on the lives of citizens," as they exploit the lapses in the political system to form and run "parallel gangster outfits" (Institute for Peace and Conflict Resolution 2002, 14). For example, both Asari and his NDPVF and Ateke and his NDVS were allegedly instrumental in the 're-election' of Peter Odili as the Governor of Rivers State in 2003 and are seen as his protégés (Naanen 2004, 9, footnote # 10). But the former soon after parted ways with him and, in order to apprehend him, the State Government declared him wanted and reinforced Ateke's gang. Okafor Ofiebor gives a helpful insight into how and why Asari parted ways with Odili: Both he and Ateke were friends and supporters of Governor Peter Odili during his re-electioneering campaign in 2003. Asari claimed that although he campaigned for Odili to be re-elected, there was an unwritten agreement that President Olusegun Obasanjo would not get an Ijaw vote since, according to him, Obasanjo does not love the Ijaw or the Niger Delta. He also alleged that Obasanjo's PDP did not win in the Niger Delta. Ironically, Rivers State recorded the highest votes cast for Obasanjo. This alarmed Asari who endorsed an advertorial on behalf of IYC, denouncing the victory allegedly declared for Obasanjo's PDP. Since the disclaimer embarrassed the Rivers State Government, he was subsequently asked to recant the press statement, but he bluntly refused. Other members of IYC were allegedly sponsored to rebut the earlier statement made by IYC and endorsed by Asari. Thus, things fell apart between them (Okafor 2006). This led to a war of attrition and battle for supremacy between the two groups between 2003 and 2004 (Naanen 2004, 7).

### *3.3.2 The Cult Dimension*

This dimension of the conflict manifests between rival cults and is usually in the form of fight for control of drug business, territory and/or oil bunkering routes. There is a plethora of cult groups in the Niger Delta, some of which have wider presence across the country. Some of the active cults include *Dey Gbam* (to be well), *Dey Well* (to be all right), *Elegam Face*, "Germans," "Icelanders," "Vikings," "Outlaws," "Bush Boys," KKK, "Greenlanders," "Mafia Lords," and "Vultures." Okechukwu Ibeanu has pointed out that these "cult groups" metamorphosed into what they are today from the previous vigilante forces and at a point took over towns like Port Harcourt, Buguma, Abonema and Ahoada. According to him, two of such cult groups stand out namely, *Dey Well* and *Dey Gbam*. Each of them identifies itself not only by a special insignia, greeting style and dressing mode, but even by the type of dry gin it drinks. Thus, while *Dey Well* drinks only Chelsea Dry Gin, *Dey Gbam* drinks Squadron Dry Gin. Many innocent people in Port Harcourt have been attacked while trying to buy Chelsea Dry Gin in areas of the city controlled by *Dey Gbam* and vice versa (Ibeanu 2004, 17).

### *3.3.3 The Economic Dimension, or Struggle for "Resource Control"*

This is the use of violent means by some youth groups in the region in their quest for fiscal autonomy. The theme of "resource control" was converted into an ideological framework within which to legitimise the conflict and endear its protagonists to the people of the region by a powerful cabal of political leaders. Daniel A. Omoweh describes the operations of this cabal as follows: "Not surprisingly, a cabal within the state not only manipulates the overall political competition in the country in its favor, but also retains control of the political ownership of all natural resources like land and oil, while excluding new entrants from access to the state's political power and access to oil and other minerals" (Omoweh 2004, 11). This group then uses unemployed idle youths as the foot soldiers in their game plan. Some of the groups are known, whereas others are not; yet, generally, the actual members remain elusive as they are fluid in their operations. Some of the known "militant" groups include: Iduwini Volunteer Force (IVF), led by 'Commander' Johnson Biboye; Egbesu Boys of Bayelsa; Ijaw Youth Congress (IYC), which is the youth wing of the Ijaw National Congress (INC); Niger Delta Vengeance Seekers (NDVS); the Chicoco Movement; and Concerned Citizens of Ika in Akwa Ibom State (CCI).

Others are the Martyrs Brigade; the Niger Delta Peace Keeping Volunteer Force (NDPKVF), which is opposed to hostage-taking; Asari's NDPVF; Ateke's NDVS; Coalition of Niger Delta Forces; and MEND, which was formed in 2006 but later split into two factions with Bayelsa and Delta State factions led by Jomo Gbomo and Godswill Tamuno, respectively. Then there is the Joint Revolutionary Council (JRC), whose spokesperson is a lady by name Cynthia Whyte. JRC comprises MEND, NDPVF and Martyrs Brigade and serves as a clearing house for the three groups. Most of these groups are well-armed and well-organised, some of them drawing membership from university graduates. Some of their modes of operation include press releases through the internet, issuing of ultimatums to the Government and MNOCs, hostage-taking, blowing-up of oil pipelines, seizure of oil company facilities, and presentation of various kinds of conditions for peace.

### *3.3.4 The Ethnic Dimension*

Nigeria is a conglomeration of about 374 "identifiable" ethnic groups, with the Hausa-Fulani, Igbo and Yoruba as the largest groups (National Population Commission and ORC Macro 2004, 1). According to Eghosa E. Osaghae and Rotimi T. Suberu, ethnicity is one of the cardinal forms of identity by which Nigerians distinguish themselves. The others are religion, region, class, gender and youth (Eghosa and Suberu 2005, 5-13). These identities have often led to conflicts at different periods in the historical development of the Nigerian state, whether the pre-colonial, colonial or post-colonial eras.

The diversity in ethnic configuration of the country is also reflected in the Niger Delta. Using language as the primary criterion of classification, Kaniye Ebeku has delineated five major clusters of ethnic groups in the region: the *Ijoid*, made up of the Ijaw-speaking people of the region and the largest ethnic group in the Delta; the *Yoruboid* Group, which is represented by three main ethnic/linguistic groups, namely, the Ilaje and Ikale in Ondo State, and the Itsekhiri in Delta State; and the *Edoid* Group, who are found mainly in Delta and Rivers States and comprises the southwest *Edoid*, which include Urhobo, Isoko, Evohwa, Okpe, and Uvbie/Effurun. The southeast *Edoid* are found in Rivers State and include Degema, Engenni, and Epie-Atissa. The fourth major group is the *Igboid*, which refers to the Igbo-speaking communities of the region who are found in Abia (Asa community), Delta (Aboh, Ika and Ukwani), Imo (Ohaji and Oguta), and in Rivers State (Ekpeye, Ogba, Ikwerre, Egbema, and Ndoni). Then there is the Delta Cross group found in central Delta: Abua, Odual, Ogoni, etc., in Rivers State and some communities found in the Lower-Cross of Rivers, Cross River and Akwa Ibom States (Ebeku 2002, 30-36).

For centuries this diversity in ethnic composition in the region has been exploited to create violent conflicts between the various groups (See Dike 1959; Jones 1963; Ikime 1969; and Alagoa and Adadonye 1972). For example, Robert McKeown, writing in 1902, says: "One town is afraid of another, and the different communities suspect one another to such a degree as never to mix, so that the difference in dialect is accentuated." Continuing, he adds, "Tribe preyed on tribe, town on town, and the white trader preyed on all to supply the markets of the New World in human merchandise" (McKeown 1902, 30-31). On their part, Ebiegberi Alagoa and Adadonye Fombo indicate that as far back as the sixteenth century the Ijaw were described as "warlike and are rarely at peace" (Alagoa and Fombo 1972, 16; cf. Omoweh 2004, 10).

The major motivation or cause of such armed conflicts among ethnic groups in the region is usually the struggle for either political supremacy (e.g., fight for power or territory) or economic control (of trade routes or trade items), or both. In his description of this scenario in the Eastern Delta in the nineteenth century, G.I. Jones notes:

> Political supremacy in the Eastern Delta could not be separated from economic supremacy. Wealth was needed to maintain favourable commercial and political relations with political leaders in the communities that controlled the up-river markets, it was also needed to buy the men who manned the trading and the war canoes, and the muskets, cannon and gunpowder with which to arm them if Bonny, Kalabari and Nembe were to maintain their monopoly of the overseas trade from which this wealth was derived (Jones 1963, 88).

Daniel Omoweh puts this into perspective when he asserts as follows:

> The Niger Delta has long had bloody confrontations: between and among the various ethnic groups, between ethnic groups and the foreign slave merchants, between the people (especially the merchant princes of the Delta area) and mercantile foreign capital as well as the colonial state, and between the people, the Nigerian state and foreign oil capital (Omoweh 2004, 10; cf. Ekeh 2004, 34).

Also, while commenting on the Warri crisis, Kayode Soremekun notes, "long before oil assumed prominence in Nigeria, there already was a phenomenon called Warri Crisis… as central as oil could be to the intensification of the crisis, this phenomenon would have obtained in the absence of oil" (Soremekun 2004, 21).

Thus, ethnicity and the struggle for economic superiority or domination among the various ethnic groups have been major issues in the region, and they predate the discovery of oil. Consequently, viewed from a

Christian perspective, if a lasting solution is to be found to the conflict, not only should the structural factors be changed, the problem of ethnicity also must be addressed so that the culture of mutual suspicion and exclusion among the ethnic groups would be transformed and replaced with a new culture of mutual trust and embrace (cf. Volf 1996).

### *3.3.5 The Communal Dimension*

Another key dimension of the militancy in the Niger Delta is communal crisis. Intra-communal conflicts are mainly caused by 'chieftaincy' and boundary disputes, quarrels over oil money from the MNOCs, cult rivalries, lawlessness and a growing culture of impunity whereby culprits go free, and/or manipulation of the youths by politicians. Several communities in the region have experienced various degrees of internal violence. Some examples are Okuru Ama, Abuloma, Amadi-Ama, Ataba, Bakana, Bukuma, Buguma, Edegbiri, Kula, Ogbakiri, Okrika, Rumuakunde, Rumuekpe, and Tombia, all in Rivers State. Intra-communal clashes have also occurred in Isoko in Delta State and in Nembe in Bayelsa State, just to mention a few. At the inter-community level, violence usually emanates from fights over land – land where there is oil deposit, as in the case of Egweama versus Liama in Bayelsa State (see Oyadongha 2007); farmland, as in the case of Bodo versus Mogho in Gokana in Ogoni, Rivers State (Onah 2007); or land used as fishing settlements, as in the fight between Akassa and Koluoama in Bayelsa State.

### *3.3.6 Outright Criminality*

It was indicated earlier that the violence in the Niger Delta could partly be explained as a hijacking and adoption of a popular struggle for resource control as an ideological framework by a group of leaders and youths for the furtherance of their personal political and economic goals. Within the struggle itself, a further mutation occurred in which some people turned it into a money-making venture and an opportunity to commit a number of unlawful acts. Some of the criminal acts include kidnapping, armed robbery, indiscriminate shooting, even at innocent civilians, destruction of property, both private and public, use of illicit drugs, illegal bunkering, piracy, membership in outlawed cult groups, political hooliganism, extortion of money from people through illegal levies, extra-judicial killings, and rape. Some examples of these acts of criminality will suffice. Sometime in September 2006, Ateke Tom, the leader of the Niger Delta Vigilante Service, was accused of master-minding several hostage-takings and armed robberies in the region, which led the Nigerian military to embark on a "Cordon and Search Operation" in Okochiri, Okrika, his base, aimed at arresting him. He escaped, allegedly with gun-shot wounds, but his second-in-command was killed, while several of his members were arrested.

An illustration of indiscriminate killing is what happened in Sangana and Warri Streets in Diobu, Port Harcourt on August 31, 2004 when suspected members of the NDPVF drove to a restaurant along Warri Street in a dark red Mercedes Benz car and shot indiscriminately at the customers, killing four persons instantly (HRW 2004, 15). An example of extortion in the name of levy is what occurred in Uvwie Local Government Area of Delta State on December 26, 2006 when one Chief E.I. Omasheye and his family were attacked by some Ekpan youths who subjected them to various unpleasant ordeals. The reason for the attack was his refusal to pay an illegal development levy of twenty thousand naira (₦20,000) imposed by the youths. During the attack the chief and some of his children suffered various bodily injuries (Ogoigbe 2007). All these indicate the criminal dimension of the conflict.

However, it is pertinent to note that all the above perspectives and dimensions are interwoven. The Rivers State Truth and Reconciliation Commission highlights this when its Chairman remarks as follows: "the problems of Rivers State rest principally on governance, politics, cultism, chieftaincy and insurgency. These happen to be pervading ills of the state. *It is to be noted however that all these five are interwoven*" (Eso 2009, 40).

## 3.4 Appraisal of the Various Perspectives to the Niger Delta Conflict

Having analysed the various perspectives to Nigeria's political economy of oil and conflict, as well as the various shades of militancy among youth groups in the Niger Delta, the next challenge is to appraise them. Their weaknesses and, therefore, the basis of their failure both on the short term and the long term, are not far-fetched. The Federal Government/MNOC perspective is the foundation of the problem and is responsible for what Festus Iyayi calls "the paradox of Nigeria's underdevelopment," or oil dilemma, against "the background of huge oil earnings" (Iyayi 2005, 1).

On his part, Okey Ibeanu describes the dilemma with three paradoxes as it applies to the Niger Delta – the paradox of plenty, the paradox of national security and the paradox of development (Ibeanu 2002/2003, 16-36). Iyayi succinctly explains this paradox as follows: "while we are one of the largest crude oil producers in the world, earning over US$350 billion between 1960 and 2000, we are a giant among the fifteen poorest nations on earth" (Iyayi 2005, 1). The oil dilemma has also being variously described as *resource curse*, or "*oil syndrome*" (Bienen 1988, 227), and "*Dutch disease.*" "Resource curse" refers to "the way a dependence on natural resources always skews a country's politics and investment and educational priorities, so that everything revolves around who controls the oil and who gets how much from it – not how to compete, innovate, and produce real products for real markets" (Friedman 2006, 31). *"Dutch disease"* is an expression used by economists to refer to "a process whereby new discoveries or favourable price changes in one sector of the economy – for example petroleum – cause distress in other areas – for example, agriculture or manufacturing" (Ebeku 2002, 99, foot note 34). But rather than call it "Dutch disease," Emmanuel Nnadozie chooses to depict it as "Nigerian disease," which he describes as an economic disorder arising from "irrational macroeconomic policies" of the government in the light of heavy inflow of oil revenue into the country in the 1970s (Nnadozie 1995, viii).

Furthermore, this perspective is mainly profit-oriented as it tends to place financial gain over people. This profit-orientation is influenced by the various economic paradigms identified. The paradigms themselves emanate from two major attitudes on the value of mineral resources (Humphreys 1983, 169-176). The first attitude, which is represented by the Federal Government, sees minerals as national wealth and focuses its energy primarily on "the 'macro' context" of such deposits;" while the second attitude sees mineral deposits as sources of income and its essence is the belief that "ore bodies are worthless until they are recognized as such by exploration, and their potential value is realized only when they are turned into operating mines" (Humphreys, 1983, 171-172). This second attitude is demonstrated by the MNOCs. The relationship between the value placed on mineral resource and how it is exploited is captured by David Humphreys when he says, "The value placed on mineral resources substantially determines the manner in which they are used, and differences in imputed values accordingly give rise to differences in prescriptions for exploitation" (Humphreys 1983, 171-172). Thus, from the perspective of the Nigerian government and the MNOCs, since crude oil is respectively viewed as national wealth and source of income, the primary drive is profit-making, even if it is to the detriment of the environment and the local communities.

In addition, Iyayi and M. Fleshman have respectively indicated that this perspective is fraught with patronage, corruption and lack of transparency (Iyayi 2005, 6-8; Fleshman 2000, 180, 186 and 189-190). Brooks notes that patronage, which is defined as "a system of rule in which all governmental authority and the corresponding economic rights tend to be treated as privately appropriated economic advantages and where governmental powers and the associated advantage are treated as private rights," is a major bane in Nigeria's political economy of oil (Brooks 2004, 140).

Evidence for corruption and lack of transparency in Nigeria's oil industry is not far-fetched. For example, in August 2007, Hamman Tukur, the Chairman of the Revenue Mobilisation Allocation and Fiscal Commission (RMAFC), the federal body statutorily responsible for allocating and disbursing federally collected revenue, reported that the NNPC could not account for the sum of ₦555 billion from the Federation Account between December 2004 and April 2007. Continuing, he adds that "the computation, means of payment and stakeholders involved with fuel subsidy are beset with lack of transparency and accountability" (Subair and Akinmade 2007; Fabowale and Anumihe 2007; and Abubakar and Anuforo 2007). An incident that illustrates the involvement of the MNOCs in the corruption in the oil industry is a report by the United States Justice Department in which it alleged that Wilbros, the US oil services company, bribed officials of the NNPC, the ruling People's Democratic Party (PDP), and the Nigerian government with millions of US dollars so as to secure the US$387 million contract for the Eastern Gas Gathering System project (Subair 2007).

Moreover, Okechukwu Ibeanu deems this perspective to be "exclusivist, regimenting, hierarchical and exploitative" (Ibeanu 2002, 164-165). For example, the NDDC that was formed as an interventionist developmental agency of state has been adjudged elitist and incapable of addressing the real problems of the region. Kaniye Ebeku offers some critical appraisal of the Commission and notes that its greatest problem is its composition since its enabling Act lacks some appropriate and necessary participatory provisions. Specifically, he indicates that the Act lacks provision for the representation of the local people, neither is there a provision for their participation in the planning and execution of projects (Ebeku 2004, 87).

Another criticism of this perspective is that the Federal Government, which is its chief purveyor, tends to focus on the direct, visible aspects of the conflict, such as kidnapping and destruction of oil facilities, while ignoring the structural and systemic factors that create them. Some of the structural factors include marginalisation, the local parlance for economic or political neglect (Zalik 2004, 401-402), and the laws that regulate the oil industry. Nnimmo Bassey, in a conversation with this author in 2009, opines that a major implication of this perspective, if continued, is the elongation of the vicious circle of violence in the region, with its multiplier effect on the rest of the nation, and the sustenance of the political, legal and economic structures that regulate the oil industry, most of which are at the root of the problem. Continuing, he argues that if these structural and cultural factors of the conflict are not properly addressed, the seeming peace ushered in by the amnesty programme of the Federal Government would be short-lived, and a more deadly conflict may replace it.

The perspective of the political and community leaders in the region merely scratches the problem at the surface without dealing with the root-causes; so it is not capable of redressing the fundamental issues of eco-social injustice associated with the oil industry. Also, it is alleged that the "elected" politicians of the region generally lack credibility as a result of election rigging and "kleptocracy" (See Manby 2004, 1-2). Such politicians often collaborate with agents of the Federal Government and the MNOCs in the prevalent culture of corruption and are accused of being the ones that organise and arm youths for their political gains (Naanen 2004, 4).

A major criticism of the non-violent civil society groups is that they are often driven by pecuniary motives regarding oil revenue and focus their attack on the Government and the oil companies, while almost completely ignoring the issues of greed and inter-ethnic and/or intra-communal biases and conflicts within the region. For example, Energy Information Administration (EIA) reports that some of the violence in Warri, Delta State, in 2004 between Ijaw groups was partly due to the struggle over which of them will receive the compensation for an oil spill that occurred in 2003. Similarly, Oguine states that "the situation (in the Niger Delta) has been complicated even further by bitter internecine disputes between various ethnic groups in the area. These ethnic disputes have very ancient roots but are aggravated by contemporary struggles for bigger slices of political power and economic resources" (Oguine 2006). Moreover, Ikelegbe, J. Makumbe (1998), A.S. Patterson (1998), and L. Diamond (1997) have pointed out certain organisational and ideological limitations of African civil society which are applicable to the civil society groups in the Niger Delta. They include problems of poverty, corruption, nepotism, parochialism, ethnocentrism, and co-optation; lack of autonomy and self-sustainability; absence of internal democracy within the groups; and ingrained self- and/or group-interests that engender inter-group conflicts along ethnic, regional, cultural and/or sectional lines. Moreover, "Civil society," argues Timothy Longman, "is a site of contestation between classes" in which they compete for political and economic power

(Longman 2010, 24). According to Ikelegbe, these factors "undermine the capacity and potential of civil society and reduce its effectiveness" (Ikelegbe 2001b, 5-6).

The perspective represented by the various youth groups is self-destructive both on the youths themselves and the region as a whole, since some of their activities boomerang on the people and the environment. For example, when they blow up oil pipelines, the oil that is spilled contributes further to the degradation of the environment and the violence they adopt leads to the loss of innocent lives and the destruction of property and social infrastructure. P.C. Nwilo and O.T. Badejo have shown that oil spills in Nigeria are caused by several factors, such as corrosion (50%), oil production activities (21%), and sabotage (28%). These spills lead to the contamination of marine shorelines and the destruction of marine life in the estuarine zone (Nwilo and Badejo 2005). In addition, the late Senator Fred Brume once reported that out of the 524 cases of pipeline rupture in 1999, 497 of them were as a result of acts of vandalism. Also, as legitimate as some of their motives and demands may be, such as the quest for true fiscal federalism, environmental sustainability and employment opportunities for the local people, this perspective is also infested with corruption, greed, parochialism, violent competition for access to resources, and ethnic rivalry as most members of the groups are part of the complex system of state patronage and the grid of political "Godfatherism," or "the big man, small boy syndrome" (Haynes 1996, 9) within what Goran Hyden aptly describes as "the economy of affection." By "the economy of affection," he refers to "situations in which cultivating personal relations is an important part of how individuals behave and make choices." According to him, this thrives in the African continent where people seek each other out for mutually beneficial reasons (Hyden 2009). Ikelegbe highlights these weaknesses when he writes as follows:

> Some civil society groups are becoming champions of ethnic sentiments, and are defining new goals that are basically parochial. The activities of the groups examined reflect ethnic militancy, regional and ethnic nationalism and separatist tendencies. The method of the groups is basically violent. Particularly evident is the fact that the groups have taken up arms in confrontation with one another and the state (Ikelegbe 2001b, 18).

Thus, the youth perspective, despite its achievements as earlier indicated, and its stance for the people of the region, is tainted with the same basic moral flaws that characterise the perspectives of the Federal Government and the MNOCs on the one hand, and the state and community leaders in the region, on the other. This is the point made by

Louis Odion when he says that the incidence of bunkering, bank-breaking, piracy, hostage-taking, or ransom-asking, is a corruption of the essence of the Niger Delta struggle for justice and fairness (Odion 2007).

Having delineated and critiqued Nigeria's political economy of oil and conflict from primary and secondary sources, the remaining chapters of this book will dwell on the data from the case studies as it investigates Baptist and Pentecostal perspectives to the crisis. The data will then be used in the last chapter to assess Gifford's claims on African evangelical Christianity and propose a political theology for Nigeria's oil economy.

# CHAPTER FOUR

# DESCRIPTION OF CASE STUDY SITES

## 4.1 Introduction

The aim of this chapter is to document and describe the case study settings using ethnographic data gathered in the course of field research in 2006 and 2007. This will then prepare the ground for the analysis and interpretation that follow in subsequent chapters. The churches selected for the research are First Baptist Church, Buguma City; Faith Baptist Church, Port Harcourt; and the International Christian Centre, Port Harcourt.

As indicated in Chapter One, three criteria informed their selection. The first is the theological: they must be churches that fall under the evangelical and Pentecostal milieu in the classification of African Christianity so as to facilitate an insider assessment of an outsider judgement on such churches regarding social witness. The second criterion is the geographical-social: it was deemed that such congregations must be ones that are located within the theatre of oil and conflict in the Niger Delta and therefore must have had direct experience of the violence. The third condition is the practical: they should be churches that are willing and accessible for the research within the time-limit of the study, especially in the light of the fact that the violence was still ongoing when the field research was conducted.

## 4.2 First Baptist Church, Buguma City

Buguma City is the headquarters of the Kalabari Kingdom as well as of the Asari-Toru Local Government Area of Rivers State. The Kalabari are a distinct part of the Ijaw-speaking peoples of the Niger Delta and live in twenty two villages and three major towns in the mangrove swamps of the eastern part of the region (Horton 1962, 197). The three major cities are Bakana, Abonnema, and Buguma, which were settled in 1881, 1882 and 1884, respectively. According to the King of Kalabari Kingdom, or the *Amanyanabo*, Prof. T. J. T. Princewill, the Kalabari formerly lived in Elem-Ama, or Old Calabar, from where they later migrated to their various towns in different parts of the eastern Niger Delta. According to another Kalabari, the late Rev. C. T. T. George, they originated from seven different places. George did not give the names of the seven places but G. I. Jones confirms a heterogeneous origin of the Kalabari in his *From Slaves to Palm Oil: Slave Trade and Palm Oil Trade in the Bight of Biafra.*

The city is not connected to the public power supply but depends on a giant generator supplied by the Local Government Council for light, mainly during the night between 6pm and 12midnight. According to Suku Ngiangia, the Pastor of First Baptist Church, Buguma City (First Baptist), the Asari-Toru Local Government Council fuels and maintains the generator and the light generated is rationed among the different sections of the city. But, quite significantly, the three big churches are permanently connected to it every night. They are the First Baptist Church, the African Church, and the Anglican Church, all of which are located along the King Amachree Road.

First Baptist Church, Buguma is located at the centre of the city, about fifty metres from the town square. According to the Church Pastor, there is no officially documented history of the church, but it is generally believed that the church was founded in 1893 by one Rev. William Hughes from the United Kingdom who was a one-time missionary to Congo and was the Director of the African Training Institute in Colwyn Bay, North Wales from 1889 to 1912. While serving as the Director of the Institute, Hughes made missionary visits to Africa, including Buguma, where he started the First Baptist Church (King 1986, 2-3; Batubo 1988, 260-261). It is the oldest Baptist Church in the region and it was from here that several other Baptist churches were planted in Rivers, Bayelsa, Cross River, and Akwa Ibom States, including Igbo land. A.K. Nodu corroborates this when he writes, "Baptist gospel was better enhanced by … Buguma indigenes who were predominantly traders. These Buguma traders helped to propagate and spread the Gospel and mission work in all places they carried on their trade" (Nodu 2004, 12).

The church auditorium is a five hundred-seat building located at the heart of the city along the King Amachree Road. It is made of cement blocks and has seven entrances and is hemmed in on three sides by residential buildings. The church choir and the worship personnel normally process in and out through one of the doors behind the rostrum. The podium is elevated by about two feet from the floor of the building. Since there is no public power supply during the day, the church runs a generator to meet its energy needs during worship and other activities. As such it is well-lit with several four-foot fluorescent tubes hanging on the ceiling. There is also a large number of ceiling fans in the auditorium. The church has a big over-head bell located outside the building to the right from the road. The bell is rung prior to church services to remind members of times of worship.

The Sunday visits took place in August 2006 and August and September 2007. The timing of the visits was deliberately planned so as to meet the church while they were already in worship. On arrival at 10:22am on August 27, 2006 the service was just beginning and it was obvious that that was the church's Baptist Students Fellowship (BSF) Sunday and, as such, the worship was led and coordinated by students. The BSF is a fellowship of students in both secondary and post secondary institutions. A Sunday is dedicated to them each year during which they lead in the worship. The Church Pastor was not present as he was said to be away to another congregation to conduct a baptismal service. The church theme for the year was boldly written on a banner facing the congregation: "My Year of Total Freedom," with Isaiah 9:4 as the theme text. The Pastor subsequently explained the rationale for the theme in the following words:

> Over the years, as I assumed my pastoral ministry in Buguma, I discovered that the church, the entire community of Buguma, the Kalabari kingdom and the Niger Delta region had been under bondage. These are spiritual, social and economic. The people of the Niger Delta region were grossly marginalised over the years. In the distribution of the nation's wealth, of which greater percentage comes from the region, the people were greatly neglected or marginalised. Total freedom from this marginalisation will help us to speak out and fight for our right peacefully. The people living in this area thus need total freedom. That is, freedom from idolatry, immorality, cultism; it has to be socially, economically and spiritually. There has to be freedom of worship. These are some of the areas the people are under bondage. Within this context, God is the only one who can bring about this total freedom using human instrumentality. God in his own time through our prayers will install a government that will look into the plight of the people.

The first song that was sung by the congregation during the 'Praise and Worship' session was an Igbo song, *"I me e, I me la, I me la, Chineke, I me la"* (translated as "Thank you, thank you, thank you, God, thank you"). Prayer of adoration was carried out mainly in Kalabari language and it was participatory. They adored God for his glory, honour, provision, care and protection, especially in the light of the violence in the city. Prayer for peace in the city, the region and Nigeria as a whole was a dominant theme in other prayer sessions. During the time of collection of tithes and offering the Kalabari Choir, mainly made up of women, rendered a special song in the local language. The Church Secretary read out the "Greetings and Announcements" in English language and it was translated into Kalabari by a female interpreter. Thus the conduct of the worship was bilingual. There was significant singing and dancing, especially by the women and youths, as the worshipers expressed themselves in moods of joyfulness, excitement and thankfulness to God.

Singing, which was the dominant item in the worship, was carried out in various forms. Out of the eighteen items on the programme, the singing of "Praise and Worship," hymns, and special numbers by the Kalabari Choir and the BSF Choir had six slots, or one-third of the number of items. Both Kalabari and English songs were sung during the session of praise. Two of such Kalabari songs which were quite popular in the church are as follows:

| | |
|---|---|
| *Oyewa oboma galema (2x)* | Let us praise him on high (2x) |
| *sotoru bee fini-ame* | The birds in the sky |
| *en lubodibiya, i ye fiari* | never labour yet they eat |

And

| | |
|---|---|
| *Jezos ani-iburoma bo* | Jesus is my Saviour |
| *ajike biya* | I will not be moved |
| *ori werisobiaajike biya* | He will not leave me alone |
| *toru akaka dibisime-semebra* | like the tree planted by the river side |
| *ajike biya* | I will not be moved. |

They also sang the following choruses:

Our Father in heaven
We glorify your name
We bow down before you

And

He's the King of kings
He's the Lord of lords;
His name is Jesus, Jesus, Jesus,
J-e-s-u-s
He's the King

The songs above were repeated several times. The first one is a collective invitation to praise God who is "on high." The emphasis is more attributive than geographical. That is, to the members it underscores transcendence as an attribute of God. Yet, despite his transcendence, he is immanent as seen in his provision for the birds. So, if he provides for the birds which do not labour, there is a guarantee that he would also provide for his people. Thus everybody should rise up and praise him. The second song begins on a declarative note as the worshippers pronounce him as their Saviour. Also, he does not abandon his own. Therefore they will not be daunted or shaken in their faith. This second song still draws lesson from the natural world: in the collective declaration of individual faith each singer compares himself/herself to the tree planted by the river which draws nutrients and nourishment from the soil. The third chorus re-echoes ideas in the first one by referring to God as Father whose abode is in heaven whom they glorify and worship. The last song reiterates the Christological focus of the second chorus by declaring Jesus to be both "King of kings" and "Lord of lords." On this occasion only one hymn from the *Baptist Hymnal* was sung, namely, "Bringing in the Sheaves."

The singing was followed by five separate prayer items. Three sets of offerings were collected, namely, the regular Tithes and Offering, Special Thanksgiving offering, and Operation Reach All (ORA)/BSF offering for mission. Also, during the worship, the Royal Ambassadors of the church, which is an organisation for boys between ages ten and twenty five, and the family of a late elderly member came forward for a special thanksgiving.

The single most dominant item in terms of duration was the sermon, which was preached by Mr. Godswill Princewill. It was titled, "The Barren Fig Tree," with texts taken from John 15:16 and Luke 13:6-9. The preacher likened an unfruitful Christian, particularly a BSF member, to the barren fig tree cursed by Jesus in Luke 13. He says, "Any Christian, or student, who is unproductive in God's kingdom work is under a curse. Just as the fig tree that bore no fruit was cursed by Jesus, such a believer is under the yoke of a curse. Therefore, every Christian student ought to be productive, especially in bearing witness for Christ." He identified two major areas of Christian fruitfulness as fruit of character and fruit of new converts. He defined fruit of character as the fruit of the Holy Spirit listed by Apostle Paul in Galatians 5:23-24: love, joy, peace, patience, gentleness, goodness, faith, meekness, and temperance. He declares as follows:

> As Christians, Jesus expects us to live lives that portray his character and to make new converts. The reason he cursed that tree was because it did not bear any fruit, even though it was full of leaves. In the same way, any child of God that does not bear fruit risks the danger of living under a curse. And I want to tell you that there are two kinds of fruits we must bear. The first, as I have already said, is by portraying in our lives that we are his followers. The qualities that prove that are listed in Galatians 5:23-

> 24 by Apostle Paul. Let's open to the passage (most members who came to church with their Bibles complied with the instruction). You see, the qualities are stated there. They are love, joy, peace, patience, gentleness, goodness, faith, meekness, and temperance. The second kind of fruit-bearing is that of making new converts by bearing witness for Christ. Christ expects us to preach the Gospel. Students, you can do this among your friends so that many more souls will be won for the Lord.

Most of the worshippers listened to the preacher with rapt attention, while a handful fell asleep. Those sleeping were intermittently woken up by ushers who were quietly moving about on the aisles to check noise making, or wake up those who were sleeping by tapping them on the shoulder. At the end of the message, the preacher extended invitations to those who wished to make various decisions, especially salvation, rededication and reconciliation, to which twenty persons responded. At the end of the worship session, members greeted one another in small groups, after which they dispersed and walked home.

However, by the follow-up visits of August 2007, the church order of activities on Sundays had changed. Hitherto, Sunday programmes began with the Sunday School at 9 am and was followed by the worship proper which began at 10 am. But on this occasion, it was sandwiched in-between the service. Also, the church had called an Associate Pastor in the person of Stanley J. Diamond. The Church Pastor Ngiangia explains the reason for the change in Sunday order of activities:

> We discovered that most of the members of the church do not attend the Sunday School. So, in order to encourage them to attend, we changed the order of activities on Sundays whereby the worship begins at 8am; we pause mid-way after the sermon for the Sunday School. Thereafter, after about an hour, we resume the worship service during which we receive announcements and collect offerings and observe any special programmes such as thanksgivings and family outing services.

Upon arrival, the Sunday School session was already on. The classes were grouped mainly along gender and/or age lines. Each class had a teacher and the teachers comprised both males and females. The size of the classes ranged from twelve to thirty members. The lessons were drawn from the Annual Sunday School Lessons as prepared by the Sunday School Division of the Nigerian Baptist Convention. According to the Pastor, more members now participate in the Sunday School than was the case previously. His message of the day is encapsulated in a one-paragraph write-up at the back of the church bulletin under the caption, "Pastor's Corner." It is titled, "The Human Nature Depends on God," in which he writes, "Creation asserts our utter dependence upon God. All that we are and have is from Him, our every breath is quite literally his gift. Our proper response is therefore true humility before him, both explicitly in our acts of worship and implicitly in the whole spirit of our living."

The worship resumed after the Sunday School, with about eighty percent of the worshippers still present. As usual, the worship was conducted in Kalabari and English. As was the case in 2006, the church theme for 2007 was boldly written on a banner and placed on the wall of the church auditorium facing the congregation. It is "2007: My Year of Wonders," with the text taken from Daniel 4:3. Approximately seventy per cent of the worshippers were women.

The attire of the worshippers was predominantly traditional: for the women it was mainly double wrapper with blouse made of various fabrics, including lace and gingham and madras cotton cloths against a variety of background colours, such as indigo, black, brown, yellow, red, and violet. The gingham and madras cloths are known among the Kalabari as *injiri* and in the rest of southern Nigeria as "George." Tonye Erekosima and Joanne Eicher indicate the cultural significance of these cloths among the Kalabari when they state, "Wrappers made from such cloth are generally worn during ceremonial occasions, and women also wear them as fashionable attire" (Erekosima and Eicher 1981, 48). There were also a few *pelete-bite* (or "cut thread") and *fimate-bite* ("pulled thread") wrappers (cf. Perami 1982, 80-81; and Daly, Eicher and Erekosima 1986, 48-51, 83). Most of the men wore *woko* and *etibo*, with a few wearing casual trousers and shirts. *Pelete-bite* and *fimate-bite* wrappers are so important to the Kalabari that when a Kalabari person wears them, "he or she exemplifies cultural authentication" (Erekosima and Eicher 1981, 51). The Pastor and his Associate were the only ones on English suit. Thus, to the congregants at First Baptist, Sunday worship is a time of cerebration and a unique opportunity for the expression of their cultural identity through fashion.

Some of the key highlights of the second aspect of the worship include welcome and announcements that lasted for twenty-six minutes and the collection of tithes and offering. There were three sets of choirs on this occasion: the main choir with 14 members (7 boys and 7 girls), the children choir made up of 6 children, and the Kalabari Choir composed of thirty women on local attire of wrapper and blouse of various sheds and colours. The colours of the robes of the first two choirs were deep red and purple, respectively.

While the tithes and offerings were being collected, the Kalabari Choir rendered some tunes in the local language, accompanied with various local musical instruments played by some of the women and a group of boys. Some of the instruments include *kuku*, an acoustic instrument in the form of a clay pot; *nkro*, a big carved hollow wood; *kuku fariye*, sticks used in playing the *kuku*; and *alili*, small two-sided local drums. Others are *bera fariye*, two carved sticks that are beaten together to generate sound; *kpokpo*, small carved hollow wood; and *alili fari-sin*, two small sticks used to play the *alili*. The vocalists and the instrumentalists played and sang in ¾ rhythm extolling the greatness and bountiful favours of God. This was accompanied with clapping of hands and rhythmic bodily movements by the singers.

In addition to the regular tithes and offering, a special offering for the poor and needy was collected. Also a special church building offering was collected. The collection of the offerings was preceded by the congregational singing of the hymn, "Bringing in the Sheaves," which is from *Baptist Hymnal* 432. Coincidentally, this was the same hymn that was sung by the congregation during the visit of August 2006.

The membership of the church is about 960, with an average weekly Sunday morning worship attendance of 635. The church operates several small groups, including fellowships for the various sexes and age groups and twenty-four "Cell Groups" spread across the city. The twenty-four cell groups all bear names from the Bible. Some examples are "Good News Cell," "Conquering Cell," "Shepherd Cell," "Victory Cell," "Redemption Cell," "Redeemed Cell," "The Master's Cell," "Providence Cell," "Endurance Cell," "Agape Cell," "Together with Christ Cell," "Royal Cell," "Divine Favour Cell," and "Triumphant Cell." The venues of the cell meetings for each week are listed on the church bulletin every Sunday. Women and children are more in number than the men. For example, the record of Sunday worship attendance on Sunday September 10, 2007 shows that there were 315 women and 220 children, as against 192 men.

Worthy of note at the August 31st 2006 meeting is the fact that the discussion was preceded by their Church-in-Conference meeting. Unlike what obtains on Sundays during worship wherein the service is conducted primarily in English and interpreted in Kalabari, the proceedings of the meeting were conducted in Kalabari, even though the minutes were recorded in English. A major matter on which decision was taken by the church was on a proposal to restructure the Nigerian Baptist Convention, the national body of most Baptists in Nigeria. Overwhelmingly the members voted to oppose the proposal. The two major grounds on which the proposal was rejected were that it is contrary to Baptist tradition and that it could lead to an infringement of the autonomy of the local churches since it indirectly alludes to possible posting of pastors. The discussion that preceded the decision was carried out in an open and frank manner as both those who stood against it and those for it freely expressed their opinions.

Although the church is located in a village setting, its ethnic composition is diverse. Thirteen different ethnic groups/dialects are represented in the church membership. Two of the three 'majority' ethnic groups in Nigeria are represented, namely, Igbo and Yoruba. Other ethnic groups and/or dialects found in the church are Kalabari, Ijaw, Efik, Ogoni, Urhobo, Abua, Ekpeye, Ikwerre, Andoni, Nembe and Okrika, with the Kalabari in the majority.

## 4.3 Faith Baptist Church, Port Harcourt

Faith Baptist Church, Port Harcourt (Faith Baptist) is an urban congregation that meets at 1 Rumuoparaeli Street, Old Government Reservation Area (GRA), Port Harcourt. According to its pioneer pastor, the late Rev. Gunn Dimorgu, the church was started in 1964 partly to make use of the piece of land that was acquired by American Baptist missionaries in 1955 and to forestall further encroachment by the Roman Catholics that had a parcel of land beside it. It has a membership of about one thousand two hundred people drawn from the various age groups, with youths and middle-aged people in the majority and is a member of the W.A. Amakiri Baptist Association, the Rivers Baptist Conference and the Nigerian Baptist Convention. All the social strata are also represented. There are at least twenty one different ethnic groups and dialects within the church membership. They include Igbo, Yoruba, Bini, Kalabari, Ibibio, Ijaw, Efik, Ogoni, Urhobo, Abua, Ekpeye, Ikwerre, Opobo, Ogba, Engenni, Bonny, Etche, Anang, and Isoko. It sponsors six other local Baptist congregations, three in Port Harcourt and three in Ogoni.

There are some weekly and monthly programmes organised by the church, such as the Wednesday prayer and Bible Study meeting, known as the "Breakthrough Hour," and the "Prophetic Morning Dew," which holds on the first day of every month. Both attract fairly good attendance by the members. The recurring themes that appear in the teachings and prayers at the two meetings are peace in the region, power and victory over satanic forces, prosperity, deliverance, job opportunities, journey mercies, healing and total wellbeing.

The church operates twenty seven House Fellowships that meet in members' homes across the city on Sunday evenings. Just as in the case of First Baptist, the House Fellowships in Faith Baptist derive their names from themes drawn from the Bible, such as "Agape," "Apple of God's

Eye," "Bread of Life," "More than Conquerors," "The Elect," "Evangel," "Shalom," "Mustard Seed," "Royal Priesthood," "Holy Lamb," "Overcomers," and "Outreach," and their venues, as well, are printed on the church bulletin. Also, there are several organisations in the church for the members. They are grouped according to gender and/or age and include the following: the Baptist Men's Fellowship (BMF), for young men and men from the age of twenty five and above; the Women's Missionary Society (WMS), for mothers and women from the age of 25 or from marriage; the Youth Fellowship, for young men and ladies between ages 18 and 35; the Baptist Students' Fellowship (BSF), for students in secondary and tertiary institutions; the Lydia Auxiliary, for unmarried young ladies from the age of seventeen; the Royal Ambassadors (R.A.), for young boys between 10 and 24; the Girls' Auxiliary (G.A.), for girls of age 10 to 17; and the Sunbeam Band, for children from age four to nine. The organisations meet during the week, either on Mondays or Saturdays. It also has a widow's ministry and runs a Nursery/Primary School and a Secondary School, the Faith Baptist College. In addition, the church operates an adult literacy programme.

A major feature in the life of the church is the use of modern technology, especially in its worship and music, which agrees with David Martin's point that "Evangelical religion has a special affinity with modern technology, because it receives messages and can amplify them" (Martin 1996; cf. Ojo 2001, 4). The Music Department is equipped with both local and western instruments such as drums, electronic keyboards, trumpet, and saxophone. There is a computerised overhead projector from which the major items of worship and wordings of songs are projected on a screen facing the congregation. In addition, the church main worship auditorium is air-conditioned.

One of the Sundays in which the church was observed during worship was July 8, 2007. The church's theme for the year was boldly displayed on a large banner on the rectangular arch overlooking the rostrum and facing the congregation. It is captioned "The Year of Unlimited Favour," with text taken from Exodus 3:21. Normally the church holds two services on Sunday mornings – 8:00am to 10:30am and 11:30am to 1:00pm. On this day the first service started at 8:05am with about a quarter of the worshippers as the rest came late. There are six main doors into the auditorium and four from the two vestries beside the rostrum. Ushers were stationed at the six doors and they directed people to the wooden pews arranged in three rows. About one thousand people attended the first service while two hundred worshippers came for the second. The Sunday School was sandwiched in-between the two services.

There was a mixture of western and traditional attire among the members. The married women appeared mostly in various hues of "up and down" wrapper and a variety of colourful blouses and head gears. Some of the men came in suits, but most of them wore local attires such as *woko*, *etibo*, and caftan of various fabrics. There was no particular seating arrangement along the lines of gender or age as members were led in to particular seats by the ushers who were strategically positioned within the auditorium to guide worshippers to their seats. The exception to this was nursing mothers who sat with their babies on the pews close to the back door. Also the auditorium was well-lit with four-foot fluorescent bulbs hanging from the ceiling. There are several ceiling and wall fans inside the auditorium. While the church is connected to the public power supply, this is generally unreliable. For this reason, the church normally uses its 27KVA generator during worship services. The church baptistry is directly behind the choir stand which sits at the back of the pastors and worship personnel. There is a gallery toward the back of the auditorium that seats about one hundred and fifty persons.

The typical Sunday worship order of the church is divided into five major parts: *Preparation* – organ prelude, processional, welcome, call to worship, and intercessory prayers; *Celebration*, comprising sharing of testimonies, hymn singing and 'Praise Worship'; *Preaching*, made up of Scripture Reading, special song by the Choir, and the Sermon, variously tagged 'the logos', 'the rhema', 'the word', or 'the message'; *Response*, comprising invitation for decisions and prayer of commitment, pastoral prayer, and tithes and offering; and *Departure*, which is made up of announcements and recognition of guests by the Church Secretary, closing prayer and benediction (usually by the church pastor), recessional and organ postlude. On average, the items that normally take the greatest share of the time are Praise and Worship Choruses (20min), sermon (45mins on average), and invitation for decision/prayer of commitment (15mins).

The sermon of the day was preached by the Church Associate Pastor in charge of Outreach and Prayer, Evangelist Abeke. He took his text from Num. 33:50-56 and titled the message as "No Compromise with the Devil." It lasted for twenty minutes. He started by saying, "The devil is a bad devil, and today I've come to announce to you that there is no compromise with the devil. You cannot afford to compromise with him because he is a wicked creature." Following a verse by verse, literal explanation of the text, he continued:

> If the Israelites are to possess or inherit the land promised them by God, they must do three things: First, they must drive out the inhabitants of the land; therefore we must know that we are in a battle and we must fight.

> Second, they were to destroy their sacred stones as they are indications of idolatry; they are signs of ancient covenants. Therefore, you must make sure that they have no place in your life. Until you destroy them, you'll not inherit the land as there are forces, satanic influence. And third, they were to dispossess the land and possess it: if you do not destroy what God says you should destroy it would become an irritant and a thorn in your life. They will become harassment to you. Whatever sickness that is meant to be destroyed in your life but which you tolerate, will terminate your life. Therefore you should not allow it.

As he preached, he moved back and forth on the rostrum and sometimes screamed and/or punched the air to emphasize some points. After the sermon, he spent about fifteen minutes inviting people to make various decisions, such as salvation and rededication, and to encourage them to pray against all forces of darkness. The congregation then stood up and joined in a very boisterous and vociferous prayer session, binding and casting every form of satanic stronghold militating against their progress, including demons, witches, principalities and powers. Thereafter, they spent some time interceding for the church, the state and the nation.

During the intercession, the congregation was led to pray for the State, the Nation, and the Church. They prayed specifically for peace in Rivers State. The ongoing militancy was mentioned, especially the dimension of kidnapping of children, including toddlers, as they besought God's intervention in the conflict. Elders and chiefs in the region were not left out in the prayers, as the congregants interceded on their behalf, that they may be God-fearing and abhor selfishness. Also, there was prayer for the newly inaugurated Yar'Adua-led Federal Government, that it might fulfil its promises to redress the past neglect of the region. In addition, the members prayed against any form of religious war in the nation and for religious leaders to be careful in their utterances so as not to incite the masses. The prayer for the church was for its spread so that churches would be planted everywhere. While this was going on, the Praise Team was singing softly at the background the following song:

> Holy Spirit, do your own thing now
> Do your own thing now
> Holy Spirit, do your own thing now
> Do your own thing now

The other major items of worship included Testimonies, Praise and Worship, Offertory, and Choir Special Rendition. Four members declared how God delivered them from various dangers: a lady testified of how God delivered her from a gang of thieves that specialised in snatching handsets from people; a girl shared a similar experience; the church organist narrated how he was delivered from a group of cultists who attempted to kill him when he inadvertently ran into them (evidence of this was on his forehead which had stitches over a deep cut); and a young architect narrated how God delivered him from the hands of some heavily armed gangs when he ran into their gun battle on July 1 that year. The following is the written account of the last testimony by the architect:

> The Lord's Day, which turned out to be my death day, if not for God's intervention! A day that my dream would have been shattered! Oh July 1, my breath would have been taken away from me and my head blown off. I would have become history on *The Daily Sun* of 4th of July (RIVERS OF BLOOD). There wouldn't have been a future or dream again. July 1, how can I forget you? I woke up on that fateful day. It was a day for 'Morning Dew'. I felt I shouldn't be there because I did not have much on me, but a friend came and encouraged me to go for the programme. While praying, Rev. Abeke mentioned that God had revealed it to him that there was a young man that was being tormented by the spirit of death and that it was in connection with a car. He said that God was ready to avert it, therefore we should pray. So we prayed and he pronounced some prophetic words of prayer and said that the enemies have been frustrated and that testimonies were on the way. At about 7:30pm I was at Emenike Street. Suddenly, a taxi cab made a u-turn and, because of the unusual turn of this cab, I and other drivers in my front decided to make use of the next junction to the right. But before we could do that, the taxi cab that had overtaken us suddenly stopped and two young men came down from it armed with heavy guns. I have never seen such guns before except with military men or in films. All of a sudden, they started shooting at us and two others started shooting at passengers alighting from a bus by my left. One of them came, looked at me and shot at me. Before I knew it, bullets had pierced through both the left and right sides into my car. There was smoke everywhere and my car was galloping because of the blast of the gun shots. Nothing was working again as the engine shut down; so the car was stationary in the midst of the shooting. The only place that was not touched by bullets was the driver's seat. At that point I didn't know what I was doing. Beside me (i.e., on the street) were dead bodies lying in the pool of their own blood. The driver of the car in my front had already been shot dead; so I was expecting mine. Suddenly my car moved and climbed the dead bodies and hit another parked car. I got up from under the steering where God hid me and started running. One of them chased after me with a gun and I fell into the gutter and he left. Then I started running down to my house. As I ran I saw dead bodies on the road. At about 8: 00pm I got to my house and called Isaac who then called Rev. Ibude to explain how God saved me from gunmen ... I don't know how he saved me. It was often said during our revival programme in June that "my bones

> will not break," and indeed God kept His word. Two days later I discovered Psalm 91 which says that because HE loves me, I WILL RESCUE HIM.

The testimony was followed by spontaneous shouts of 'Praise the Lord' from the members; some clapped their hands; while a few blew whistles. They then sang the following choruses,

> We are saying thank you Jesus,
> Thank you our Lord;
> We are saying thank you Jesus
> Thank you our God.
>
> You are good, you are great
> O Mighty God;
> You are good, you are great
> O Mighty God.

And the Igbo song,

| | |
|---|---|
| *Ekele diri Chineke* | Thanks be to God |
| *Ekele diri Nwa nya* | Thanks be to his Son |
| *Ekele diri Monso* | Thanks be to the Holy Spirit |
| *Onye neme nma* | The One that does good |

The feeling that came across from the members as a result of the testimony was that of appreciation and gratitude to God for his deliverance. Also, it confirmed to them the efficacy of the 'Morning Dew Programme' and the trustworthiness of the prophetic utterance of Rev. Abeke in the morning of that fateful day.

At the end of the first service, the church entered into the Sunday School period during which members went to their various classes. Meanwhile, towards the end, those for the second worship service arrived and also joined in the Sunday School. The same order of worship was used in the second service.

The message of the Associate Pastor on Sunday August 15 was entitled "Our God is in Control," with text taken from Psalm 46. It lasted 45 minutes. An excerpt from the sermon is as follows:

> When we say that God is mighty, why do we say he's mighty? He is mighty so that we would be mighty too. God is powerful so that we might be powerful too. God is rich not just for himself but also for his people, that they might be rich too. God is a present help in the midst of trouble.

> God is in control; God is in charge; God is fighting for you; God is a fighter. Therefore do not be afraid.

On the implication this has for believers, he says, "Nobody among us will be put down; nobody among us will be put to shame; no one among us shall be forsaken." To each of these declarations the congregation responded with loud shouts of "Amen." The same pattern of response continued throughout the period of the sermon. Also, in the course of the message, a good number of members spontaneously got up from their seats and placed offerings inside wooden boxes placed on the aisles. In concluding the message, Abeke said, "Each circumstance has an expiry date. God uses our cases to shine. He uses our circumstances, our problems, to brag, to prove himself to the world. Only be sure that your anchor is holding on him."

## 4.4 Hilltop International Christian Centre, Port Harcourt

The Hilltop International Christian Centre, Port Harcourt, a.k.a. "Oarhe World Outreach," describes itself as "a charismatic ministry/church for today and the future." It was started on June 6, 1993 at the Oroworuko Town Hall in Port Harcourt where it remained until 2000 when it moved to its permanent site. Its main auditorium is a modern structure that can seat up to three thousand worshipers in the main seating area, the two galleries and the overflow. It shares boundaries with the St John's Campus of the Rivers State College of Education on one axis, the railroad track by the Garrison Junction on the other, and the Nwaja Creek in the Trans-Amadi Industrial Layout within the Orochiri Layout.

Originally, the church was known as "International Christian Outreach Centre" before it changed its name to "The Hilltop International Christian Centre," based on Matt. 5:14b. Although they assert that each of these names has its significance, they explain the change this way:

> The Lord instructed that we pursue a ministry to the nations from Port Harcourt with an outreach focus driven by the love of Christ. This informed the incorporation of the word 'outreach' in the name. This focus has not changed because a house set on a hill is placed there to be revealed – to reach out. Thus, 'THE HILLTOP' is a powerful replica for 'outreach'.

They describe the impetus that led to its establishment as follows:

> Countless numbers of people today feel confused and disillusioned. Many people have lost hope. Broken relationships, failed marriages, youth suicides, drug abuse, rising crime rates and social dilemmas are sobering

> indications of the hopelessness that many are experiencing. Some do not know who they are or where they are going. Unemployment exacerbates these problems. Many feel misunderstood, excluded or oppressed. Often a lack of cultural understanding hinders harmonious relationships in our diverse and pluralistic society.

Having identified the above problems affecting the society, the church set out to respond to them by raising men and women who are equipped to understand the contemporary times and who are filled with Christ-like compassion to serve the hurting and confused. Also, it desires to train men and women who seek to live the Christian life so as to make impact in the moral life of the society, people who can take up key leadership roles and influence the society with gospel values, and people who will inspire and give hope to others.

Regarding its teachings, Hilltop groups them into five areas, namely, Vision and Mission, Aims and Objectives, Statements of Faith, Core Values, and Strategy. In seeking to carry out its mission of turning "the sinner into a saint," the church states its Vision Statement thus: "To equip believers to reach the un-reached by building Christ-centred relationships which respond to people's hurts, needs and interests. We operate home cells to enhance this pursuit." This emphasis on outreach is further seen in its motto, "Reaching the nations with Christ's love." Its central aim is "to help raise a people who will focus on the Lord Jesus Christ." This aim is under-girded by the following six objectives: testing of everything against the background of the Scriptures, Christ-centeredness in every area of life, soundness of mind and high Christian standard in morality, ministry orientation, faith, and spiritual empowerment. The Statement of Faith of the church covers eight major themes: the Word of God, the Trinity, sin, salvation, Christology, the Holy Spirit, eschatology, and the Church.

The core values of the church are exaltation, evangelism, edification, and extension. These values are required of every member of the church as it stipulates that as "a child of God, particularly for any member or intending member of the Hilltop, you must be soaked in (them)." Exaltation is carried out through corporate and private worship in which members recognise that they are purchased with a price through the atoning death of Christ on the cross. Through evangelism the church seeks to place emphasis on the Great Commission by bringing lost people to God by word and deeds of love. Since it believes that "the body of Christ is…composed of interdependent members, each having received from the Holy Spirit gifts which are to be exercised," the purpose of edification is the mutual building up of one another through Bible teaching, discipleship training, personal evangelism, personal commitment and encouragement.

Extension is the last core value through which it seeks to reach out to the wider society with the compassion and justice of God's Kingdom. Thus, the church declares: "The church can never settle for its comfort in a world wrecked by problems ranging from moral decay to political and financial corruption and oppression – violence, crime, unjust use of resources and wealth acquisition, distribution and other social scourges that require answer."

In order to actualise these core values, the church has a seven-step strategy: Step one is to build a relationship:

> Rather than crudely begin with the message of Christ, we seek to make connections so that when we do share the message, we have credibility. People are more inclined to consider a message from a trusted friend than from a stranger. The love and interest we show (will) cultivate the ground in preparation for sowing the seeds of the gospel.

The second step follows the first one and it is for members to verbally share the gospel, since "it is not enough just to be nice people. Unbelievers must learn why we are the way we are and who can save them and how." The rest of the steps are: three, invite an unbeliever to service; four, become a part of God's family; five, participate in a home cell; six, serve in the body of Christ; and seven, be a good steward of personal resources as they are bestowals from God.

From an initial membership of seven, the church has grown to a membership of about one thousand five hundred, with an average weekly worship attendance of about one thousand two hundred. Its male/female ratio is about equal, with slight aggregate numerical edge by the females. The same pattern is repeated during the mid-week services of the church. Each Wednesday the church organises faith-building worship service in which an average of four hundred worshipers attend. A major feature of both the Sunday and Wednesday worship services is the attendance of new comers. On average thirty four new comers attend worship on Sundays and three on Wednesdays.

The aspect of the church demography with the sharpest contrast is the age of the members. By far youths (i.e., those between the ages of eighteen and thirty five) constitute the majority of the membership. This may not be unconnected with the overall tone and direction of the church's ministry and emphasis. Apart from its worship programmes, the church has an active Media Ministries Department through which it runs two television broadcasts in Port Harcourt. The TV programmes are "New Experience," began in 1999 and presented by Chris Oarhe, the Senior Pastor, and "Heart-to-Heart" (H2H) started in 2001 and anchored by Stephanie Oarhe,

Co-Pastor and wife of the Senior Pastor. H2H is a major social ministry of the church whose objectives are stated as follows:

> To see married couples happy, take their relationship to higher heights, drive it to deeper depths and really find fulfilment in their homes. Also through H2H, singles are educated on how to achieve the kind of relationship God desires for them. Single parents are helped to overcome the challenge of bringing up their children. Heart-to-Heart aims at bringing about positive changes in the lives of many people... In addition to all of these, H2H has a foundation, a non-governmental and non-profit organization set up with the primary focus of feeding, clothing and caring for hungry, malnourished and sick children; aiding women and children to improve their quality of life; creating awareness on how teenagers can handle their emotion and eradicate teenage pregnancy; and educate people on AIDS. All these are done through various projects and avenues that will bring about this development as contained in the H2H brochure of information.

In order to achieve the above aims, H2H creates

> a relaxed atmosphere where issues revolving around these set of people are discussed openly, transparently and genuinely in order to produce lasting solutions. Such relaxed atmosphere includes live talk shows which are subsequently aired on the television for the general public, seminars and programmes organized in secondary school, tertiary institutions, etc.

Several youths have been drawn to the church through the programme. And just as in the other two churches in this study, Hilltop is a multi-ethnic congregation, drawing membership from twenty different ethnic groups.

Apart from ministering to the spiritual needs of its members and visitors, the church has a Welfare/Benevolence ministry to both members and non-members. Through the ministry it gives out clothing, foodstuff, and money to those in need on a regular basis. It also ministers to those that are shut-in in hospitals and prisons through its Hospital/Prison Ministry Team. Also, in collaboration with the NGO, NDDC/AM Projects it regularly carries out free medical care outreach programmes to rural communities in the region. Educationally, it owns and runs a nursery/primary, the International Nursery/Primary School.

Moreover, it takes care of the pastoral needs of its members through its Outreach (Pastoral Care) Fellowships spread across Port Harcourt. There are seventeen of such groups and each group has a leader and an assistant. It has also opened branches in Oyigbo, a suburb of Port Harcourt, and in South Africa.

Unlike the two other churches in this study, Hilltop operates a top/down leadership structure. There are two structures as explained by the church, namely, the Administrative Structure and the Spiritual Structure. According to its Administrative Structure, the flow of authority begins from the President who is the Chairman, Board of Trustees and the Chief Executive Officer. Next to the President is the Executive Vice President. Other officers are the Executive Secretary, who is the Chief Administrative Officer; Regional Pastors who exercise spiritual oversight over regions; Directors who oversee directorates and departments; Mission Administrator, who carries out administrative oversight at the International Headquarters; Church Administrator, who oversees the administrative needs of a region; District Administrative Officer in charge of a district; and Church Administrative Officer who oversees a local church.

Its spiritual structure is clear from the following sketch supplied by the church:

Figure 1 Spiritual Structure of Hilltop

The Lord Jesus Christ
↓
The Senior Pastor
↓
The Co-Pastor
↓
Associate Pastors
↓
Other Pastors
↓
Deacons
↓
Ministers
↓
Church Workers

Just as in its administrative structure, the spiritual flow of authority follows a top-down pattern. According to them, at the top of the hierarchy is the Lord Jesus Christ. Next is the Senior Pastor who is assisted by the Co-Pastor, who incidentally is the wife of the Senior Pastor. Associate Pastors are next in rank to the Co-Pastor. After the Associate Pastors are the Deacons. The second to the last cadre are the Ministers, which refers to those involved in the various ministries of the church such as Evangelism,

Music, and Media. The cadre of Church Workers is the lowest and refers to all other voluntary officers of the church who serve in various areas of the church life and ministry.

However, the above hierarchical structure notwithstanding, Hilltop is egalitarian in its ministry and in the relationship among members. This emanates from two major theological grounds: their belief in the equality of every believer in Christ before God following personal conversion experience and the free distribution of the "gifts of the Holy Spirit" on members by the Holy Spirit, which enables every member to actively and responsibly participate in the life, growth and development of the church. This active participation is expressed by the members through the various departments and ministries in the church. In all there are twelve departments, each of which supervises a number of ministries. The Departments are: Hilltop Connect, Evangelism and Publicity, Media, Heart2Heart, Special Ministries, Couples' Fellowship, Christian Growth Series (CGS), Life Savers, Music and Kingdom Arts, Youth and Singles, Hilltop Service, and Prayer and Intercession.

One of the Sundays in which the church was observed during worship was on August 5, 2007, which incidentally was the church's "Anointing Service" day. There were many cars of different makes parked in the car park located within the church main compound and an adjacent parcel of land. Some of the cars had stickers on them with the following words: "Empowered to Succeed," "Anointed for Exploits," "On the Wings of the Eagle," "There shall be no loss," and "From Glory to Glory." J.K. Olupona notes this feature in Pentecostal church life when he says, "Car stickers and T-shirts, on which are inscribed important messages of the group such as 'Jesus Is Lord', 'Are You Saved?' and 'The Blood of Jesus', adorn their homes, cars and places of work" (Olupona 2002, 14). The church generator was on and the auditorium, which is air-conditioned, was almost filled to capacity with worshippers. Church ushers were strategically located at the main doors and within the auditorium to lead people to their seats. The church choir, known as the "Praise Ambassadors," wore a uniform of white tops and black trousers, or skirts, and sat beside the rostrum facing the congregation to the left. The worship leaders on their part sat on the other side of the podium facing the choir. Most of the pastors and worship leaders wore suits. The Senior Pastor walked in during the course of the worship as the praise session led by the Praise Team was going on.

The singing was accompanied with dancing and shouting as "Music and dance constitute an important dimension in the prolific mass mediation of religion among the Pentecostals" (Kalu 2008, 122). The

Praise and Worship session lasted for thirty minutes, after which a young mother came forward to share testimony of how God miraculously healed her baby of fluid in the head and other medical complications despite the negative report of medical doctors on her case. Several members placed bottles of 'anointing oil' beside the pulpit. Two prominent choruses sung during the Praise and Worship include:

> O Lord, my God, When I in awesome wonder
> Consider all the works Thy hands have made,
> I see the stars, I hear the rolling thunder
> Thy power throughout the universe displayed
> Then sings my soul, my Saviour God to Thee
> How great Thou art, how great Thou art
> Then sings my soul, my Saviour God to Thee
> How great Thou art, how great Thou art

And

> J-e-e-e-sus, J-e-e-e-sus
> Your name is a miracle
> Your name is a comforter
> Your name is a wonderful
> J-e-e-e-sus

The highlight of the worship was the message, which was brought by the Senior Pastor. He took his main text from Matt. 12:33-37 and his sermon was entitled "The Law of Confession." The Scripture reading was done by all as each worshiper was directed to read out the passage from his/her own Bible version. The Pastor began the message by declaring thus:

> You are not an ordinary human being; you are the righteousness of God. You are unstoppable; you cannot fail. You cannot be defeated. That sickness is too small to kill you. Your blood cannot be polluted. HIV cannot kill you. You cannot be poor. That house is too small for you. They find out what may happen; we decide what should happen.

On his approach to preaching he says, "I am not a sermoniser. Here we expose the word." Going further, he defined confession as "making a statement that aligns with what God says" and then delineated four types of confession in the Bible: the confession of a sinner, the confession of a believer of his sins, the confession of Jesus as Lord, and the believer's confession of the word of God. In order to reassure the members and build their faith, he went on to say,

> I can assure you that after this meeting there will be a new story about you. No man can determine your future apart from you. I hear the Lord telling me to tell his children that you have remained in this land for too long. This anointing service will change you. It will reflect in your job, life, clothing, (and) house. You cannot be barren; no Christian can be barren. You must come to a point in your life where you know that 'God is able' becomes 'I am able'. Every time you say, 'I cannot,' God calls it an evil report. When you say, 'I am sick', God calls it an evil report: Num. 13:32. Whatever you do not have in your *word*, you cannot have in your *world*.

Citing Dan. 3 and 2 Cor. 4, he continued thus:

> You don't just have physical eye; you also have spiritual eye. Your role is to believe in your heart and confess with your mouth and leave the rest to God. The problem with many people is that they're stopping God from acting for them. Therefore this message is intended to stop you from stopping God.

While preaching, the pastor moved back and forth on the rostrum and some times climbed down the steps to the worshippers. Also, during the course of the message, members spontaneously rose from their seats and presented offerings in bags kept by the rostrum. Some screamed and danced, while others jumped up and one or two ran up and down the aisles. The sermon lasted for one hour fifteen minutes (9:50-11:05am)

Anointing of worshippers with olive oil immediately followed the preaching. The Pastor introduced the anointing by saying, "God is pouring out liquid fire through this oil. No demon can withstand it," to which the congregation shouted a very loud "Halleluiah." Thereafter a group of officers came forward to receive bowls of 'anointing oil' from the pastor for sharing among the congregants. They passed round the bowls into which each worshipper dipped fingers to collect some drops. While this was going on, the Praise Team led in the singing of some worship songs, especially

> Halleluiah, Halleluiah, Halleluiah
> Praise the Lord (2xs)
> My Jesus conquered Satan,
> Halleluiah, Praise the Lord
>
> He's Lord; He's Lord, Amen
> He has risen from the dead, He's Lord
> Every knee shall bow, every tongue confess
> That Jesus Christ is Lord!

And

> Anointing, fall on me;
> Anointing, fall on me;
> Like the power of the Holy Ghost,
> Anointing, fall on me!

When everybody had collected and the bowls were returned, the Pastor then asked each of them to place their hands on their head. He then led them in a special prayer.

From the statements and attitudes of members toward the anointing, it appears they give it certain significations in terms of its purpose: for healing, dedication or consecration, empowerment for breakthrough, and special unction for protection and victory over the forces of evil. Also, it deepens "the appropriation of the Pauline teaching on the charisma and charismata connecting these to the fivefold ministry" (Kalu 2008, 116). This should be compared with the three usages of anointing oil identified by Brian Doak and William Griffin, namely, healing, consecration, initiation; approval; and physical protection (See Doak Griffin 2006, 340-341). The following testimony by a member of the church illustrates one of the above significations:

> I ate yam one morning, which caused me stomach upset. Not just any type! This particular one was terrible. I couldn't walk upright. So I felt I won't be able to come to church. But I practically dragged myself down to church. When pastor anointed me, the pain left me immediately. Now I can stand uprightly. I thank God for His Divine healing!

The anointing with oil was followed by the collection of two sets of offerings: the regular tithes and offering and Mission Fund offering meant for the support of their mission efforts in some rural villages in the region. The service thereafter came to a close with prayer and benediction and members converged in small groups to exchange pleasantries before they dispersed. Meanwhile, before going, a good number of members went to the church media stand to buy or place order for some Christian literature on display and/or the audio/video tape of the day's message. Some of the books in the church book shop include *Pursuit of Purpose* and *Glory of Living*, by Myles Munroe; *Only Believe*, by Smith Wigglesworth; *Goal*, by Bryan Tracy; and *Managing Your Emotions*, *by* Joyce Meyer. They also include *Praying with Your Pastor* by the Senior Pastor, and *Flying on the Wings of Relationship* by one of the Associate Pastors in the church and the Senior Pastor's brother, Austin Oarhe. Also available was the H2H magazine.

## 4.5 Comparative Perspective on the Three Churches

The three churches were founded at different epochs: First Baptist is more than a century old having been established by a British missionary in 1893; Faith Baptist was founded in 1964 through the influence of American Baptist missionaries; while Hilltop was started in 1993 as an indigenous Pentecostal church. Comparatively, Hilltop is well adapted to ministry in the region, especially among the youths and middle class, as it has more structures in place to reach out to the society beyond the walls of its auditorium than the two Baptist churches. This is mainly through its Free Rural Health Ministry in collaboration with the NDDC/AM Projects and its media outreach programmes, namely, "New Experience" and "Heart-to-Heart." The church also engages in socially relevant activities, such as road maintenance and provision of food and clothing to the less-privileged.

Whereas the two Baptist churches could be described as practising *structured spontaneity* in their worship, Hilltop practises *spontaneous structure*. That is, the worship format of the former has some formal order, yet the order is embellished with exuberant spontaneity as members actively participate in it through singing, praying, dancing, clapping and shouting (cf. Witte 2008, 699). In the case of Hilltop, spontaneity precedes structure without obliterating it. In addition, Sunday worship sessions serve as the biggest avenues for African spiritual and cultural expression among the members as seen in their music, dance, prayers, and dress. It is noteworthy that in Hilltop, unlike First Baptist and Faith Baptist, women wear trousers to church. In the two Baptist churches it is seen as culturally incongruent and biblically untenable for women to be in "men's attire."

In polity, the two Baptist churches are autonomous, congregational and democratic; whereas Hilltop is independent, hierarchical and participatory. In theological configuration, whereas Hilltop is *pentecostal-evangelical*, both Faith Baptist and First Baptist are *evangelical-charismatic*, which confirms Gifford's claim made from Ghanaian perspective that "the majority of Baptist churches now seem little different from Pentecostals" (Gifford 2001, 76). Kwame Bediako calls this process the "pentecostalisation" of "mainline churches" (Bediako 2000, 312; cf. Asamoah-Gyadu 2008, 193-194), while Gifford calls it the "charismatisation" of "the historic or mission churches" (Gifford 2004, 174; and Gifford 1994, 524). That is, whereas Hilltop is a Pentecostal church with strong evangelical emphasis, the two Baptist churches are evangelical churches with robust charismatic fervour, thereby presenting an integration of missionary Christianity with

Pentecostal theology. Furthermore, the three churches believe in the dynamic aspects of the ministry of the Holy Spirit in the life of the church.

They also manifest certain tendencies that show the footprint of Pietism in their theology. They include affirmation of the possibility of a personal experience of God, beginning with conversion, insistence that that experience should have direct impact on personal life, the requirement of a Christian community that is different and separate from the world, and Biblicism (Bundy 2002, 610-611). This pietistic leaning is not unconnected to the European Protestant missionary background of much of African Christianity (Ankrah 1977, 155-161). From an African perspective, the biblicism noted above is based on their belief that the Bible is "God's liberative, life-giving word, rooted in Christ, 'the saviour of the world'" (Okure 2000, 15).

It is this biblicism that perhaps explains the approach to biblical interpretation and preaching observed in the three churches, which is basically the same. It is an approach oriented towards biblical literalism and practical application by proffering solutions to concrete problems and one in which the main focus is the "encounter between the Biblical text and the African context" (Ukpong 2000, 3). According to Ernest van Eck, "In this method(s) the Biblical text is linked to the African context in such a way that the focus of interpretation is on the communities that receive the text rather than on those communities that created the text, or the text itself" (van Eck 2005). Ogbu Kalu calls this "bumper sticker" hermeneutics, or "experiential literalism," in which the "emphases are on the experiential, relational, emotional, oral aspects of faith; the immediacy of the text; and the freedom to interpret and appropriate the multiple meanings of the biblical texts" (Kalu 2008, 266-267). For example, consider the following words from the Associate Pastor of Faith Baptist in the message he preached to his congregation on Wednesday July 27, 2007 (his text was from Joshua 11:1-11, 16-23):

> Tonight I just want to prophesy to you that you're victorious... I have agreed with the Lord today to prophesy to you that you are victorious... May be you're looking for a job, I prophesy to you that you'll have a job; may be you are looking for something, I prophesy to you that you will have it... Whatever the Lord has said, it's all about faith... Believe God that has given you victory – victory in your job, victory in your marriage, victory in every area… Joshua conquered all areas. That means all areas of life. Whatever may be your problem, the Lord says you can conquer all the land… Life is all about fighting. I prophesy to you, 'People are moving out from Port Harcourt. But Port Harcourt shall not be deserted. Those who left shall return. I prophesy peace to Port Harcourt. Peace shall return to the city'.

Thus Gifford is apt in his description of this preaching style:

> It is immediately practical. It is directed to the climax of the evening, the 'altar-call' when all who have not yet done so are exhorted to come forward to be 'born again'. Thus the speaker's aim is not to stimulate thought or to explain ideas or even to expound the Bible. It is not to impress with his erudition. Nor is he out to draw admiring applause… His goal is to have the listener running to the front … to proclaim a public commitment… The preaching is thus an exercise to a specific goal (Gifford 1987, 67; cf. Burgess 2008, 32).

A related phenomenon that was also observed is the rate of ecclesial migration, or the change of church affiliation. This is the "tendency of believers to move from one church or organisation to another" (Marshall 2009, 133). It is usually due to a number of factors ranging from doctrine, worship, polity, to personal grievances (Gifford 2004, 170). A good number of such members were noted in Faith Baptist and Hilltop. In fact, some former members of Faith Baptist were seen in Hilltop. Also, some former members of some Pentecostal churches joined Faith Baptist during the period of the field research. For example, seven new members were formally received into the church membership on Sunday July 8, 2007. Among them were former members of Pentecostal churches. On enquiry, one of them claimed to have left his former church on three major grounds, two negative and one positive. The negative reasons he gave were that he did not like the way the pastor of his former church was running it as a personal business without accountability and that the church failed to come to his aid when he had a major personal problem. On the positive side, he said he heard of Faith Baptist through a friend and, upon coming to the church, liked the way the "word", i.e., the sermon, was preached and the openness with which it was run. Thus it has to be noted that the gravitation of change of church is not only from the missionary-established churches to the Pentecostal ones; it is both ways, but this is not often acknowledged in academic publications on African Christianity which seem to portray the phenomenon as a one-way traffic. For example, Gifford states that new charismatic churches "attract not so much non-Christians or (say) Muslims, but mainly Christians from other strands of Christianity, both Catholic and Protestant" (Gifford 2004, 170).

But this constant drive to change church affiliation portends some dangers to the Christian Church in Africa due to its orientation towards spiritual consumerism. These include shallowness in Christian discipleship and half-hearted commitment and loyalty to their respective churches in which individual members place more emphasis on their personal needs than on faithfulness and devotion. Bruce Shelley and Marshall Shelley, writing from an American perspective, succinctly capture this when they comment as follows: "Many people assume that their needs count more that their loyalty. If their needs go unmet, they are quick to switch to another church, just as they would doctors, grocery stores or airlines to find better service" (Shelley and Shelley 1992, 166). Also, the Baptist theologian, John Hammett, critiques this "consumer attitude" as one in which "Members view their commitment to their church as they view their commitment to shopping at Wal-Mart" (Hammett 2005, 157).

Having given the ethnographic description of the three congregations, the next chapters will document, describe, analyse and interpret the major categories from the data that were gathered in the course of the field research. The thematic approach in which data from the three case studies are integrated is followed in the analysis based on the fact that the three churches show remarkable similarities in doctrine, liturgy and praxis within the context of the conflict, despite the denominational differences between the two Baptist congregations and Hilltop. This approach was deemed to be the most appropriate in order to facilitate the smooth flow of discussion and thereby avoid undue repetition if the site-by-site approach were followed. Thus the common themes are selected and integrated based on their frequency, distribution and dominance as determined from word count of the data.

In order to judiciously address the first two of the major research questions of this book, the analysis of the data is subdivided into two major blocks. The first dwells mainly on the question of their theology – their beliefs and teachings – in the context of oil and conflict and is addressed in chapters five and six; while the second focuses primarily on the issue of their praxis – their practices and activities – within that context and is discussed in chapter seven. The two blocks roughly correspond to what John Parratt describes as 'orthodoxy' and 'orthopraxis', respectively (Parratt 1990, 7). Chapter Five will address the categories of God, Jesus Christ, and the Holy Spirit, while Chapter Six will dwell on the church. The penultimate chapter will focus on the categories of crude oil and conflict. The last chapter will then focus on the third objective of this study, namely, to evaluate Gifford's claims on African evangelical and 'fundamentalist' Christianity as earlier highlighted in Chapters One and Two based on the data.

# CHAPTER FIVE

# THE TRINITY: GOD, JESUS CHRIST, AND THE HOLY SPIRIT

## 5.1 Introduction

This chapter seeks to present the first part of the answer to the question of the underlying theological issues and beliefs of the three congregations in the study within the context of oil and conflict in the Niger Delta. Based on their frequency and distribution as extracted through word count from the official documents, songs, prayers, sermons, statements, and conversations held with a cross section of individual members of the three congregations, the first set of predominant theological categories that emerged in the case studies are God, Jesus Christ and the Holy Spirit. Their occurrence in the three churches are as follows: God: 16 in First Baptist, 66 in Faith Baptist, and 30 in Hilltop, giving a total of 112; Jesus Christ: 28 in First Baptist, 25 in Faith Baptist, and 23 in Hilltop, giving a total of 76; and Holy Spirit: 8 times in First Baptist, 10 in Faith Baptist, and 12 in Hilltop, giving a total of 30 times.

Due to their prominence in the worship, programmes and activities of the churches, special attention will be paid in this chapter to their songs in the description and analysis. This is mainly because it was discovered in the course of the field research that singing in various forms is the single most dominant item of worship in terms of number and variety and second only to the sermon in duration in each of the three churches. For example, in Faith Baptist, in a worship session that lasted for about 145 minutes, singing by the Church Choir and congregational singing of hymns and praise and worship songs, accompanied with dancing, took a total of 45 minutes; and in First Baptist, out of a total of eighteen items in the order of worship, singing took one-third. The same pattern was noticed in Hilltop. Thomas Oduro alludes to this when he notes that "music takes about three-quarters of the duration of worship in many Christian communities in Africa" (Oduro 2008, 92). Thus, much of the oral theology of these churches is expressed through songs as music plays certain roles in their worship. John Mbiti expresses this when he says, "Through music, singing and dancing people are able to participate emotionally and physically in the act of worship. The music and the dancing penetrate into the very being of the worshipping individuals" (Mbiti 1969, 61). The preponderance of music in their worship most likely draws from the African culture in which music and religion go hand in hand (cf. Agu 1990, 51). The other major forms of expression of oral theology of the three congregations that are of particular interest in this chapter are their sermons and prayers. These are then supplemented with their statements. The first major theme to be considered is God.

## 5.2 The Doctrine of God

The 'official' belief of the two Baptist churches in this study about God could be deduced from the Statement of Faith of Faith Baptist, in which it is declared:

> There is only one God and besides Him there is no other. He is the creator of heaven and earth and all therein. He has no beginning or end. He is all knowing, all wise and Almighty. God is personal. He has will and purpose. He is the absolute ruler of the universe, the redeemer and preserver of all His creation (cf. Nigerian Baptist Convention 1993).

On the Trinity, the document states as follows:

> God reveals Himself to us as Father, Son and Holy Spirit while continuing always to be one God. The three revelations of God as Father, Son and the Holy Spirit are a mystery which defies human understanding or ability to give adequate expression of man's experiences with his three-fold nature. In the Trinity God reveals Himself as Father, Son and Holy Spirit, each having distinct attributes but without division of nature, essence or being.

Similarly, in The Hilltop Statement of Faith, it is declared that "there is one true Holy God, eternally existing in three persons – Father, Son and Holy Spirit; each of whom possesses equally all the attributes of deity and the characteristics of personality."

The above affirmations on God and the Trinity convey certain truths on the beliefs of these churches about God. The truths centre on his nature, personality, relationship with the created order, and his role in creation. Thus, they affirm the oneness and uniqueness of God, his creation of and absolute sovereignty over the universe, his eternity, as well as his omniscience and omnipotence. He is a person who relates with the world as ruler, redeemer and preserver.

On the Trinity, the documents assert their being one-in-three, their personality, and their distinctiveness. Yet they share the same nature, essence and being. Because of the mysterious nature of the concept, the churches declare that the doctrine is not fully comprehensible for humans, whether in understanding or in expression. These statements on God and the Trinity are in consonance with the basic tenets in such historic ecumenical documents of the Christian Church as the Apostles Creed, the Nicene Creed of 325 that affirmed the full divinity of Christ, and the "Chalcedonian definition" which developed out of the Council of Chalcedon of 450 and which asserted the full and complete humanity of Christ.

These official affirmations perhaps influence the declarations made about God in the statements, prayers and songs of these churches. Thus, God is declared in various ways as *Tamuno*, *Baba*, *Papa*, *Ogologo Mma agha*, the God of Abraham, Isaac and Jacob, a Man-of-War, and the One in control. Respectively, these could be seen in the following set of songs regularly sung in the three churches:

| | |
|---|---|
| *Tamuno-be oremiye* 2x | The *Lord* is good |
| *Ibalafaa yeh* | He's good to me' |
| *Ibala fa ---* | I'm not afraid again; |
| *(Ibalafama ba ro foriyeh)* | The Lord is good, He's good to me |
| *Ibala fa ---* | I'm not afraid again |
| *(Ibalafama ba ro fori yeh)* | The Lord is good, He's good to me |
| *Ibala fa --- (2x)* | I'm not afraid again (2x) |
| | |
| *Baba ye ---* | *Father* of life, |
| *Je ki ijo-ba re de,* | Lord, let thy kingdom come, |
| *Baba ye* | Father of life |
| *je ki ijo-ba re de,* | Lord, let thy kingdom come, |
| *Ka ase re bere, Lori mi* (4x) | Let your will be done in my life (4x) |
| | |
| *Our Papa weh dey for heaven,* | Our *Father* who is in heaven |
| *Na you we sabi oh---,* | You're the One we recognise |
| *Everything for our lives,* | Everything in our lives |
| *Eh dey for your hands,* | Is in your hands |
| *Everything for our life,* | Everything in our lives |
| *Eh dey for your hands.* | Is in your hands |
| | |
| *Ogologo mma agha,* | You're the *Mighty Sword* |
| *Jehovah bu Ogologo mma agha* | Jehovah, You're the Mighty Sword |
| *'K'm nwere* | That I have |

The God of Abraham, Isaac and Jacob,
Jehovah is a Man of war
His mercies endureth forever and ever
So praise His holy Name

He's in control
He's in control
Jehovah *Nissi* (interchanged with Jehovah *Shammah* or Jehovah *Rapha*)
He's in control

The theme of 'God-is-in-control' was also expressed by the preacher at Faith Baptist on August 15, 2007, which is cited in the preceding chapter. The title of his sermon was in fact, "Our God is in Control."

God is also affirmed as an awesome God that reigns from heaven, with wisdom, power and love. The following song, which is popular in the three congregations convey this:

Our God is an awesome God, he reigns
From heaven above, with wisdom, power and love
Our God is an awesome God

Consequently, the appropriate human response to this awesome God is praise and thanksgiving as expressed in another set of songs:

| | |
|---|---|
| *Oboma gbeme,* | It is good to praise the Lord, |
| *otokini gbeme,* | It is good to worship the Lord, |
| *Sonakri na teme dabo,* | The *Creator* of heaven and earth, |
| *Oboma gbeme* | It is good to praise the Lord |

I will praise him from everlasting
Everlasting to everlasting
I will praise him form everlasting
Everlasting to everlasting

Praising the Lord, Praise him,
In his sanctuary praise him,
In the firmament of his power

You are good, you are great
O Mighty God;
You are good, you are great
O Mighty God.

And the Igbo song,

| | |
|---|---|
| *Ekele diri Chineke* | Thanks be to God |
| *Ekele diri Nwa nya* | Thanks be to his Son |
| *Ekele diri Monso* | Thanks be to the Holy Spirit |
| *Onye neme nma* | The One that does good |

There was even a theological interpretation to the root of the crisis in the region as the following submission by a member of Faith Baptist who is a retired oil company manager shows:

> The trouble, or problem, in the Niger Delta is due to the fact that God is annoyed. God, the all-knowing God, saw the Niger Delta and its problems; because he is the one that created the place, with its difficult terrain and its people, he decided to compensate them by placing oil inside the ground with the intention of the people using it to develop their harsh environment. But man in his wickedness decided to take the oil from here to another place, leaving the environment devastated. That is why God is annoyed and that's why we will never have peace until God is happy.

The member above attributes the emotion of anger to God which, according to him, is caused by human wickedness and injustice that undermines his creative benevolence on the region based on his foreknowledge.

The Pastor of First Baptist highlighted what he calls 'total freedom', which he believes is only possible through the help of God and which is highlighted in the previous chapter. Thus, while describing the rationale for and the meaning of his church theme for 2006, namely, "Total Freedom," he states that it includes freedom includes from idolatry, immorality, and cultism, and has to touch the life of the people "socially, economically and spiritually."

Not only does God bring about total freedom, he is also seen as the one that saves or delivers his people from everyday mortal dangers. This is the theological context of the testimony shared in Faith Baptist on July 8, 2007 by a young male architect narrated in chapter four above. According to his testimony, the only way he could explain his escape from the shoot-out was simply God's deliverance. This perspective was expressed in eight other testimonies of God's deliverance that were shared in Faith Baptist during the period.

But what makes this deliverance possible, especially since other people died in the same scenario wherein they experienced deliverance? The answer could be deduced from the following account by a female member of Faith Baptist who is a civil engineer as she narrates how she survived a gun battle in Buguma City, 'by God's grace', when the church went there for the funeral of a late member:

> I remember there was a time a member of the church died and as is the custom of the church, we went for the burial in Buguma. While we were there, suddenly, the crisis started and members of the church had to run helter-skelter, and some of us even ran into the swamp and slept over night. Those of us that did not run that far went into other people's houses. In short, some of us were even crawling under other people's beds. We were there for quite some time but by *God's grace*, some of us came back around 4 o'clock. As for me I was there till around 6pm. When I came out to go I heard a gunshot and there was no way to go; so I decided to take the water side path to get to the outskirt of the town. But somebody told me that that place was even worse and that if I wanted to go, that I should follow the straight road and I saw that I couldn't. Then luckily for me, *God helped me* and I escaped. That day a lot of people died but by *God's grace* none of us died but we suffered. (Emphasis added)

Furthermore, God is conceived of as the one that answers prayers. The importance of this belief could be seen in the fact that throughout the period of the fieldwork, prayer was the most prominent direct response of members of the churches to the crisis. For example, a young graduate in Hilltop says, "This church has responded to the crisis in the Niger Delta by way of prayer; several times prayers have been raised for the crisis"; while a middle-aged mother in First Baptist says a similar thing and describes the breadth of the prayer topics: "We normally pray during the crisis. Even in the church we pray for the state; we pray for the community; we pray for the country so that God should intervene in the crisis taking place in the Niger Delta."

The belief in the efficacy of prayer in resolving the conflict in the region is strongly held by members of the three churches. This is more so when it is complemented with fasting. Thus, an elderly woman in First Baptist says, "The way I look at it is that if we are to survive, we have to pray and fast for the Almighty God to help us and assist us because if we are to get what we need in the Niger Delta, or to maintain peace, it is only God that can make it for us." Similarly, a young male school teacher in the same church emphatically asserts, "When we pray, something actually happens."

The prayers are expressed in words as well as in songs. An example of the latter is the following prayer-song sung in First Baptist during one of its Sunday morning worship: "*iyoki ye ibubele ye mie, Tamuno iyoki ye teke ibubele ye mie*" – "Take me and use me as you like, God take me and use me as you like." A similar prayer-song says, "Lord prepare me, a sanctuary, pure and holy, tried and true, with thanksgiving, I'll be a living, sanctuary, for you." This was sung at both Faith Baptist and Hilltop.

There is also belief in the role of prayer in community safety as expressed in the following statements by an adult male in Hilltop:

> When my community, Amadiama, was bombarded and was at the verge of destruction, we resisted it, not with physical strength but, as a community, we went to God in prayers. We did not go with physical weapons. Yet there are people that came into our place with assorted kinds of weapons; but, when they came, they were unable to penetrate. Rather they killed themselves because we went down on our knees in prayer. I lost the only thing my grandfather left for me through destruction and looting. The Nigerian army came in and almost everything was carted away, what he left as a legacy – a library (he was a king); all those things were burnt and when we saw those things we felt bad. But what do we do? We can't challenge them because they feel they have the guns. But we went down on our knees. The church helped me in the sense that I came to my ministry – the publicity ministry – and we held hands together in agreement and I solely believe in prayers. There are sometimes I wake up 6am, there will be shooting for 24hours, I can't go out. My house is almost at a transit point for all that nonsense, but by the grace of God, flying bullets did not touch any member of my family, although there were casualties. But the rate was reduced because of the prayers and we still believe and we're still in that prayer; and … whatever force or whatever bullet they would shoot, it would not penetrate.

The key points of the above account are: first, there was an attack by two opposing gangs who fought and 'killed themselves'; second, the Nigerian army came in to stop the fighting, but ended up meting more destruction on the people and their property; third, the major response of his community was prayer; fourth, the church through his particular ministry was his source of succour; and, fifth, there is belief that prayer is a sort of bullet-proof. Beneath all these is the belief that God answers prayers.

The preceding discussion shows that God is perceived in various ways in the three churches. He is described as creator, owner, sustainer, redeemer, man of war, *Tamuno*, *Baba*, *Papa*, *Ogologo Mma agha*, punisher, fighter, avenger, helper, enabler, the one who blesses, and the one who answers prayers. Some of his attributes that were noted include omnipotence, omniscience, justice, grace, power, and peace. A further analysis of these descriptions and attributes would yield four sub-categories: absolute attributes, authoritative attributes, affable attributes, and adversarial attributes. Absolute attributes refer to those descriptions of God by the members which are unique to his nature, which John Mbiti describes as the "eternal and intrinsic attributes of God" (Mbiti 1969, 30-36); while authoritative descriptions refer to those that focus on his sovereignty over all of creation. Affable attributes are those descriptions of

God by the members that centre on his friendly disposition towards his own people that make possible personal relationship with him; while adversarial attributes describe God in relation to the enemies of his people. These are now taken up.

### *5.2.1 Absolute Attributes of God*

The attributes of God described by the members in relation to his nature and which are unique to him are omniscience, omnipotence, justice, grace and peace. First, God is regarded as all-knowing and such knowledge includes his foreknowledge of the difficult terrain of the region prior to creating it. This is the point made by the retired oil company manager in Faith Baptist cited above who asserted that God the Creator foreknew the peculiar nature of the Niger Delta environment and decided to compensate its people with crude oil so as to enable them to utilize the resources from it to develop their area.

Not only is God all-knowing, he is also all-powerful. The idea of the power of God is expressed in various ways through statements, sermons and songs. While describing her response to the loss of her relations and property to the conflict, a middle aged female pharmacist in Faith Baptist says, "In such a situation nobody can do anything. What influenced me was just that I know that I serve my God and I know that there is nothing God cannot do." The power of God is also expressed in terms of his might. This was expressed by the Associate Pastor of Faith Baptist in his sermon of July 15, 2007 titled, "Our God is in Control."

This theme of the power of God is further expressed in some popular choruses sung in Faith Baptist, such as *Ogologo mma agha*, or 'Mighty Sword'. Literally the Igbo version says, "Long fighting sword, God is the long fighting sword that I have." When asked to expatiate on this imagery of God, one of the leaders in the Music Department of the church said, "What we mean by that is that God is the one that defends and fights for us; so we depend on him to fight for us in our battles." This concept further is taken up below.

Another major attribute of God that emerged in the case studies is his justice. This belief in God's justice is very widespread in the three congregations and partly informs the views of their members regarding the root of the problem in the region. To them justice is not just a forensic theme but a moral quality related to God's omniscience. It goes this way: God created the Niger Delta environment, which is a very difficult one; and, in his foreknowledge, he 'blessed' the region with crude oil so as to compensate for its unfriendly geography, thereby balancing his creative justice. But human wickedness turned an otherwise positive endowment into an instrument of oppression against the people of the region and, in so doing, inverted God's creative justice. That inversion manifests in neglect, marginalisation, discrimination, and oppression. Since he does not support injustice and since his creative justice is being hampered by human injustice, it is believed that God will eventually intervene on behalf of the people of the region to restore justice.

The next core attribute of God alluded to in the churches is his grace. Just as on the issue of justice, divine grace within the context of the conflict is not merely a cerebral concept but a living reality that is encountered in daily experience. The female civil engineer in Faith Baptist cited earlier makes this point when she says: "... We were there for quite some time but, *by God's grace*, some of us came back around 4 o'clock... luckily for me, *God helped me and I escaped*. That day a lot of people died but, *by God's grace*, none of us died but we suffered" (Emphasis added). From her perspective, it was this grace that made it possible for some of them who ran into the swamp or other people's homes to survive unhurt, whereas some other people died. Thus, the grace of God is not just "unmerited favour" regarding the salvation of the soul but also God's help in time of need.

The last absolute attribute of God noticed is peace. He is seen as the God of peace. The word 'peace' is used so much in the church and yet no attempt is made to define it. But, on further reflection and from a summation of their contributions, it is apparent that peace for them in the context of the conflict means a state of harmonious non-violence whereby people are able to live their normal lives freely and without fear. Such a state is made possible by God and whereby it is lost, he is the one who can restore it, since he is the God of peace. When the condition of harmonious non-violence is supplanted by anarchy and brigandage, the members' faith is rattled as they struggle to balance their belief in God's ability to sustain peace and the situation that vitiates it.

Within this scenario, the believer's role is essentially to pray for "God's intervention." Such prayers for peace to be restored are couched in various forms: supplication, intercession, prophecy and 'release'. The first two forms take place on a regular basis, while the last two are occasional. An example of each of the last two forms is as follows: during the prayer meeting of Faith Baptist held on Wednesday July 11, 2007 in which the Associate Pastor led, the church spent about an hour praying for peace in the region. Towards the end of the programme, as he was concluding the session, the pastor declared: "People are moving out from Port Harcourt. But Port Harcourt shall not be deserted. Those who left shall return. *I prophesy peace to Port Harcourt*. Peace shall return to the city!" to which the congregation shouted a loud 'Amen!'

Also, following the invasion of Bane community in Ogoni where the church sponsors a preaching station, members at the prayer session were directed by the pastor to "release the peace of God to the land of Bane." This was in response to the mayhem that occured in Bane in the Khana Local Government Area of Rivers State in the early hours of Monday the 9$^{th}$ of June 2007 as a result of a fierce gun battle between two rival cult groups, *Dey Gbam* and *Dey Well*. Thus, the peace of God is not just an inward experience of the believer, but a possession which can be shared and extended to the society by means of the spoken word. After the directive, members vigorously prayed, using a variety of expressions, such as "I release the peace of God to Bane," "Peace must return to Bane," and "I bind the agents of darkness causing the violence in Bane." See below for a further analysis of their theology of prayer.

### 5.2.2 Descriptions that Portray the Authority of God

Next to the absolute attributes ascribed to God in the churches are descriptions that portray his sovereignty and rule over creation. In that regard God is conceived as *Tamuno*, Creator, Owner, Ruler, and Controller. Already these conceptions have been anticipated in the previous ones.

For example, through a very popular chorus in the three churches, God is addressed as *Tamuno*. In Kalabari worldview *Tamuno* belongs to the fourth and highest level of reality and is understood as the maker of the universe. (The other levels are *tomi kiri*, 'the place of people'; *teme*, the realm of spirits; and the level of personal *tamuno*. See Horton 1962, 199-207.) *Tamuno* is conceived as female who "formed all things in the world out of mud." According to Horton, she differs from individual creators and is spoken of as *opu tamuno*, 'the Great Creator' (Horton 1962; see Tasie1970, 7). She watches over the world, and ultimately nothing can happen in it which she has not willed (Horton 1962, 206). Thus, in Kalabari understanding, *Tamuno* is the name for God and is so used in Christian expression, but without the female gender ascription. This is because in traditional Kalabari usage *Tamuno* is addressed with feminine pronouns, whereas in traditional Christian thought masculine pronouns are used in reference to God. For example, whilst *Tamuno* is called 'mother' in Kalabari worldview, God is addressed as 'Father' in mainline Christian theology. Therefore, through the song, the church extols God for his goodness; hence they would not be afraid again.

Whether called God or *Tamuno*, the divine being is conceived of as the Creator of the world, particularly of the Niger Delta, and its oil. Not only is he the Creator of oil, he is also the Owner. Since his being the Creator of the world entitles him to its ownership, his ownership of it confers on him the position of ruler. The reign of God is an often repeated theme in the three congregations, and it is expressed in songs, statements and supplications. One of such songs is the three-line lyric earlier highlighted which describes God as awesome and one who reigns from 'heaven above' with wisdom, power and love. In other words, God's rule covers the whole of creation and is not capricious. Rather, it is governed by the above three virtues.

Whereas the reign of God is certain, still there is a "not-yet" aspect of it. This "not-yet" dimension elicits prayer for its full manifestation by the church. Such prayer is done through intercession and in songs. One of such prayers for God's kingdom to come and his will to be done in each individual's life is conveyed in the Yoruba chorus *Baba ye*, or 'Father of Life." God is addressed as *Baba ye*, the one who originates and sustains life and to whom life belongs. It is noteworthy that the ascription of the title of *Baba* to God is present in traditional religion among the Yoruba. This is attested to by Idowu in his description of the Yoruba divinity, Orisa-nla, whom he says is regarded as *Baba*, "The Father," *Ba' nla*, "Great Father," and *Ba' t'orisa*, "Father of the divinities" (Idowu 1966, 29). Thus, after calling him the father of life in the first line, the second line of the song is a cohortative supplication for his kingdom to come. The last two lines convey the direct result of the coming of the kingdom in the life of the people if God were to grant their request, namely, the doing of his will in their lives.

The other regal conception of God is that of controller which develops out of his sovereign rule; hence, he is said to be "in control." This is the belief that God is personally and actively in charge of every situation in the life of his people, even when such situations seem inexplicable and are deleterious in effect. It is a reassuring theological palliative in the midst of hopelessness and the barrage of attacks from 'the enemy', since the battle is a spiritual one. That is, the belief serves as a caveat to those who might be at the fringe of caving in to untoward circumstances that there is still hope. That is, even if they have lost it, God has not.

### 5.2.3 Affable Descriptions of God

Not only is God conceived of as awesome, he is also winsome, friendly and intimately related to his people. He is not a dictatorial, far-distant and detached sovereign but a very close and personal benefactor to those who believe in him. He is God-in-relationship-with-his-people. Therefore he is seen as Father, Helper, Enabler and the One who blesses.

The ascription of fatherhood to him is done in several ways and especially through two songs in which he is addressed as Father. The first is the one that addresses him as *Baba ye*, 'Father of life,' which has already been highlighted. The second song is the one in Pidgin English which calls him *Our Papa weh dey for heaven*, or Our Father who is in heaven. In traditional African society the father plays important roles, such as progenitor, provider and protector. Thus, he must have pedigree, power and pith if he is to be a successful father. Possession of plenitude becomes an added advantage. Uncertainty about one's paternity leads to being tagged 'an illegitimate child', or bastard, whereas having an illustrious person as father is a thing of pride. Therefore, when church members address him as 'Father of life' and 'Father who is in heaven', they are emphasising the personal relationship that exists between them. He is the source, sustainer, and supplier of life and its requirements. He is in heaven, transcendent; yet he is present in the midst of his people. This transcendent-presence elicits confidence from the people as he is the only one they know and believe in. Therefore, they are not bastards shut out from their Father's inheritance but bona fide members of his household who are entitled to the plenitude that comes from him. It is pertinent to note the two terms used to describe God as father in the songs referred to above. Respectively they are the Yoruba word *Baba* and its Igbo equivalent *Papa*. Both emphasise paternal endearment, intimacy and respect from a child/young person to his/her father or an elder for, in normal usage, both terms could be used to refer either to one's biological father or to an elderly man.

The next two affable depictions of God are inter-related. He is Helper and Enabler. Physically and emotionally he is the helper of his people. Already reference has been made to a member of Faith Baptist who attributes her escape from gun battles to the help of God through his grace. Thus, God's help is seen as an outworking of his grace. Whereas God's *help* is experienced regarding physical preservation and protection from danger, his *enabling* has to do with the strength to withstand the emotional and spiritual agonies associated with personal loss to the violence. A young lady in Faith Baptist who is a university student and whose father was killed in the conflict expresses the deep struggle she goes through, especially concerning her conviction that it is God's will for her to forgive those that killed him in the following words: "Personally I was not happy when I lost my dad and even up till now I'm praying to God to get over it.

I find it difficult to forgive them; *I'm just praying that God will enable me...* I want to forgive them because I know that it is God's will for me to forgive, even though I'm struggling with it."

The last depiction of God under this sub-category is that of the one that blesses. This theme is also very pronounced in the three congregations. He is specifically seen as the one who blessed the region with crude oil to compensate for its environmental disadvantages. Thus, a middle-aged man in First Baptist queries, "Look, we have these things and God has blessed us with them, why is it that we are not having it for our social and economic wellbeing?" This notion of God as the one that blesses has also been noted in Old Testament scholarship by Claus Westermann who contends that the God who 'saves' is also the God who 'blesses' (cf. Westermann 1979).

### *5.2.4 Adversarial Attributes of God*

The last set of descriptions of God noticeable in the three congregations is the adversarial attributes, or the roles he plays on behalf of his people and the oppressed, against their enemies and oppressors. Whereas the affable descriptions focus on God's relationship with his people and his benevolence towards them, and basically are connected to his grace, the adversarial portrayals represent his perceived antagonistic stance against the oppressors and enemies of his people, whether spiritual or physical, and develop out of his justice. They are also related to the imagery of God as 'the mighty sword' earlier mentioned. Therefore he is pictured as Deliverer, Fighter, Avenger and Punisher.

First, God is deliverer. God's deliverance from various potential misfortunes or violence is a recurring theme in the three churches. Specifically, on Sunday July 8, 2007 four members of Faith Baptist shared individual testimonies on how they were *delivered* from the hand of armed gangs and/or robbers. Two Sundays later a young mother shared testimony of how she and her family were *delivered* by God from a fire outbreak near their house that claimed the life of a child. But by far the testimony of God's deliverance that best encapsulates the members' understanding of divine deliverance is seen in the personal testimony of the young architect in Faith Baptist described in the preceding chapter. In the testimony, he narrates how his attendance at a prayer programme of the church where he received prayers and prophetic pronouncements from the members and the pastor served as the key to his narrow escape from death in the hands of militants. All of it is cast in the mould of God's deliverance from mortal danger. This is because it contains the five major ingredients in personal deliverance motif seen in most testimonies in the church. They are *danger*, *preservation*, *victims*, *escape*, and *interpretation*. 'Danger' refers to the circumstance, attack or misfortune one was faced with; 'preservation' is the outlet or mechanism of safety in the midst of the danger generally assumed to be provided by God; 'victims' refers to other persons who could not escape the danger; 'escape' is the actual flight from the danger; while 'interpretation' is the hindsight theological signification given to the whole incident as miraculous divine deliverance. Thus, deliverance does not exclusively refer to spiritual exorcism but covers every area of life.

The above description focuses on God's deliverance from danger and agents of violence, but the question arises, what is his attitude towards the perpetrators of the violence? The answer comes from the next descriptions of God that emerge. He is the God that fights, avenges and/or punishes. While reassuring the members of God's protection over their lives in one of his sermons, the Associate Pastor of Faith Baptist declares: "God is fighting for you; God is a fighter. Therefore do not be afraid." A number of theologians have also addressed this motif of God as fighter, or warrior, such as Norman Gottwald (1979), Millard Lind (1980), Lois Barret (1987), Gerhard von Rad (1991), and John Howard Yoder (1994).

If God is not punishing the enemies of his people or oppressors for their injustice, he is avenging their specific infractions against them. Thus, a pharmacist in Faith Baptist who lost her relations to the violence and who was cited earlier chose not to avenge their deaths because of her belief that vengeance is of God. The idea is this: when you suffer personal harm or loss, do not take the laws into your own hands to fight back because, as a just and reliable fighter, God will avenge for you in due time.

God is seen as 'a man of war' in relation to his covenant with his people. That is why he is called 'the God of Abraham, Isaac and Jacob', a reference to the covenant he established with the nation of Israel as represented by the three patriarchs. Yet, the theme of mercy is mixed with the idea of warfare, which demonstrates that Jehovah's way of waging war is different from the way the world does.

In the context of the Niger Delta conflict, the warfare is interpreted as a battle between two camps: Satan, his *agents*, both human and non-human, and *agencies* on one side, and God and his people on the other. *Agents* of Satan are depicted in the form of personalities, whether human, such as witches and oppressors, or non-humans, such as demons, principalities and powers. The satanic *agencies* in the warfare are such conditions or forces as sickness, poverty, and childlessness. The believer fights on the side of God against all these agents and agencies with the assurance that God is his/her 'Mighty Sword' and guarantor of victory. Against such enemies

God is fighter, punisher, and avenger. This is the general theological background to the sermon of the Associate Pastor of Faith Baptist on Sunday July 8, 2007 in which he uses martial language to say, "There is no compromise with the devil"; "the devil is a bad devil"; and "If you do not destroy what God says you should destroy it would become an irritant and a thorn in your life." It is also related to his sermon of July 15, 2007 captioned, "Our God is in Control," wherein he describes God as mighty and powerful. Thus, as the mighty sword, God is unmatchable, unassailable and invincible.

A major idea contained in the conception of God as 'Mighty Sword' is justice. Since he is the epitome of justice by virtue of his providential care of the Niger Delta by endowing it with crude oil, it is he that can bring about justice and equity to the region. The political meaning and implication of this theological underpinning of justice is taken for granted such that they do not care to articulate it, even though the idea is mentioned more than ten times in the focus group sessions. However, a juxtaposition of their views would show that justice is understood as a multidimensional concept and the ground of harmonious coexistence, the giving of due credit or honour, the allocation of fair share of resources, an attitude of "*egbe bere, ugo bere*," or "live and let live," a reflection of democratic equality, a loyalty to the law, and the keeping of faith and trust (cf. Nzomiwu 1988, 33-39). Since these ingredients of justice are lacking, they believe that God is the one that can bring them to pass. This is why an illiterate mother in First Baptist says, "*God is the only solution* who can maintain the peace for us by bringing justice." (Emphasis added)

Also, 'Mighty Sword,' in reference to God, carries the idea of personal relationship with him. This is emphasised in the song by the last phrase, "That *I* have." God relates with his people as individuals and is each believer's last resort on the ground that he/she has him, not as a possession, but as a backer, helper, enabler, and protector. This ties up neatly with the conceptions of God as *Baba* and *Papa*, which underscores the truth of God-in-relationship-with-his-people and links up with their theology of prayer.

## 5.3 Theology of Prayer

To say that prayer is the major response of the three congregations to the crisis is an understatement. It is at the heart of their life and activities – during Sunday worship, at prayer meetings and bible studies, at home and in private, as well as at ecumenical gatherings – the three churches and their members engaged in a variety of prayers – adoration, praise, thanksgiving, supplication, petition, 'binding and casting', prophecy, declarations, and especially intercession in relation to the conflict. They prayed alone and in groups; they prayed out loud, either shouting or screaming, and in silence; they prayed holding hands together and raising hands up; they prayed standing, dancing, kneeling and/or sitting; and they even prayed through songs. Regularly they added fasting to their prayers. For example, while in the field, First Baptist had a three-day scheduled prayer and fasting programme in September 2006, while Faith Baptist embarked on a similar programme for one week in July 2007·

The major prayer topics, in their order of prominence, include peace and quick resolution of the conflict in the region; divine intervention; good governance at local, state and federal levels; warfare against demons, witches and principalities and powers; safety on the roads and in homes; deliverance, breakthrough, success in exams and in businesses; promotion; and the safe delivery of pregnant women. Thus, the breadth of their prayer points is wide ranging. Beneath all these activities of prayers is a strong belief in the God that answers prayers, since he is "in control" and "in charge."

A significant perspective to prayer was also manifested by some militants in Buguma and a member of Hilltop, namely, prayer as a sort of bullet proof or talisman to ward off enemy bullets. This view among the militants was expressed when a group of them went to one of the pastors in this study to demand that he should pray for them so that the enemies' bullets will not hit them. The pastor's account of the episode is as follows:

> One day a group of ten well-armed youths – armed with assorted weapons and automatic guns such as AK 47, machine guns and dynamites – came to me. I was in my house upstairs and the main door downstairs – made of steel – was locked. These boys came and began to bang at the door, calling for the pastor. My wife and several members of the church were with me. We were all frightened. As the boys persisted in their banging, I decided to go down and see them. My wife protested but I insisted that I must go since we didn't know why they were calling me. When I got down and opened the door, they requested that I should pray for them so that no evil would befall them as they fight against their enemies. I took them to the church auditorium wherein was gathered a large number of people who had run there for safety. I entered the church, while the boys remained outside with their guns. I simply prayed, 'God, let your will be done'. While I was praying the boys heard that the opposing gang had come to attack; so they left.

This understanding of prayer as a sort of bullet-proof shield by the militants derives from a worldview that is basically theological and a spirituality that is all-embracing. Specifically, it originates from the *Egbesu* warrior ethos in which it is believed that a person could receive some shielding from physical harm by receiving some concoctions and undergoing some spiritual rituals performed by the priests of the god. Martha G. Anderson calls the concoctions "bulletproofing medicines" which include charms, herbal concoctions, gourds (*atu*), belts (variously known as *egbe*, *gbinye*, and *ideri*), and other items. Once taken, they believe that they have potency to render them invisible and enable them to seek out or ensnare and annihilate their enemies (Anderson 2002, 96-97). Caroline Ifeka throws more light to this belief and practice as follows:

> *Egbesu's* warrior youth priests are initiated into secret knowledge of 'medicines' conferring immunity to enemy bullets; initiates are supported by 'mothers of the community', by male elders in their customary role of the ancestors' representatives, as well as by priests of shrines to gods of the sea, mangrove swamps and the fertile soils of drier upland areas (Ifeka 2006, 723-724).

From a Christian point of view, as expressed by a member of Hilltop, prayer replaces the concoctions and the paraphernalia of rituals performed by *Egbesu* priests.

Putting together all of the above, certain points emerge on the members' theology of prayer in relation to oil and violence as prayer plays seven crucial roles in their lives, namely, theological, spiritual, political, physical, emotional, communal, and social. Theologically, prayer is seen as an acknowledgement of human weakness in contrast to an omnipotent and omniscient God. Also, it is a form of statement expressing a number of attitudes toward God, such as dependence, faith, and soliciting of divine intervention. This belief in the omnipotence of God is expressed in various ways through sermons, songs, prayers, and statements. Thus, he is described as *ogologo mma agha*, fighter, deliverer, liberator, avenger, and punisher. Not only that, by praying to him, they also express their implicit faith in him as being able and willing to act and therefore are soliciting his demonstration of that willingness with real action. This belief is expressed in various ways through stickers with such words as, "God is in control" and "My miracle is on the way."

The problem with this is that prayer sometimes serves as an implicit excuse for human inaction. This is due to the fact that at times it is a tacit disavowal and/or avoidance of human personal and/or collective responsibility and role in confronting or, at least, engaging with the perceived unjust structures of state. This is usually couched in the language of "we need divine intervention." Nigerian Catholic theologian

Matthew Hassan Kuka raises this same issue when he notes and describes this tendency among Nigerians to believe that all that is needed is prayer as the "*Siddon Look*" ('sit and look on') syndrome, a phrase coined by the late Nigerian Minister of Justice and former Governor of the old Oyo State, Bola Ige (Kuka 2000, 10).

In a situation of poverty, illiteracy, and general feeling of impotence, it is understandable why people would cast a near absolute role to prayer almost to the neglect of human responsibility. Also, in a situation where basic amenities are lacking, where social infrastructure is virtually comatose, where unemployment rate is very high and the few jobs available are available only to the well-connected, people depend on God for all manner of breakthroughs. Thus, a parent would pray for her child to pass his/her examination, yet during much of the school calendar teachers are on strike and so the children are not taught. Even when the school is in session, some parents are unable to buy the required books for their children and when some of the children return from school, they have to assist their parents in the shop or hawk goods on the streets and so have very limited time for adequate preparation for exams. Yet, they seriously pray to God for good results for their children.

The same thing applies to prayer for journey mercies. It is a known fact that most roads in the region are bad and that a good number of the vehicles plying the roads are not road worthy. This is compounded by the fact that most drivers, especially commercial vehicle drivers, are not properly trained. Yet, when they travel or their relation is about to travel, they seriously pray for journey mercies. This absolute dependence on God for virtually everything fills the vacuum created in the life of the people by the failure of the state to provide basic amenities. Therefore prayer for the people is not just a theological expression but also a political statement: through the spirituality of prayer, the people inadvertently indict the Government for its failure.

Also, to the people, prayer plays a crucial spiritual role. It serves as a means of penetrating the diffused layers of wicked spiritual forces and thereby reins in their evil powers and subjects them to the power in the name and blood of Jesus and to the desires of the ones praying. The wicked forces dealt with include demons, witches and wizards, as well as the personified powers behind such misfortunes as 'premature' death, failure (in marriage, business, and studies), accidents, and childlessness. Simply put, prayer to them is a weapon of warfare given to the believer in Christ to overcome the spiritual hosts of darkness. To them prayer serves as an existential moment when the powerless tap into power and thereby seek to order the affairs of the powerful through supernatural power. This is seen in the two types of prayer observed in Faith Baptist, namely, prayer as prophecy and prayer as release. In the former the members make declarative statements concerning a situation which they believe have the efficacy to bring about the desired result; while in the latter they presume to be the agency of change via their words.

Politically, prayer is understood by members of the three churches as a dynamic means of connecting to the realm of the powers, or of communicating with God, with a view to bringing to bear on the political realm his will, or, as in a lot of cases, the desires of the believers. Thus, prayer is not just a means of communicating with and/or surrendering to God but also assumed to be a sort of political tool in the hand of the believer for the enforcement of divine will in a given situation.

The physical role of prayers could be deduced from the bullet proof role attached to it by the member of Hilltop who testified how he believes that it was the prayers of his community and his church that protected them from the bullets of both the Nigerian army and the militants. Thus it serves as a talisman, or 'bullet proof', to ward off evil.

Emotionally speaking, prayer serves as a means of succour for grief and loss and as an outlet of emotional catharsis within the context of anarchy, fear and violence. Several members lost their loved ones, businesses, houses and other property to the conflict. In the face of the confusion occasioned by calamity, prayer becomes a channel of coherence as they seek to grapple with the aftermath of each episode of the crisis. In addition, prayer plays a communal role among the members of the churches and the wider Christian community as it is an expression of communal solidarity and a means of mutual support and encouragement.

Finally, prayer plays a social role in the life of the people. Through it the church seeks divine intervention for peace in the wider society by way of intercession. In fact a male youth in First Baptist states it this way, "If not for the prayer of the churches, the conflict would have been far worse than it is."

## 5.4 Jesus Christ

This section seeks to describe and document the answers from the case studies to the question of Christology –what they have written and/or are saying, praying and singing about Jesus within the context of oil and conflict. Or simply, who is Jesus to them in the light of Nigeria's political economy of oil and conflict? As in the preceding section on God, the answers are gathered from their official documents, songs, prayers, and statements. Hilltop declares in its Statement of Faith as follows:

> Jesus Christ is the eternal second person of the trinity who became human by miraculous conception and virgin birth. He lived a perfect life in obedience to the Father and voluntarily atoned for the sins of all creation by dying on the cross as their substitute, thus satisfying divine justice. He rose from the dead in the same body, that same body glorified in which he lived and died. He ascended into heaven and sat down at the right hand of the Father where He, the only mediator between God and man, continually makes intercession for His own. He shall come again to earth, personally and visibly to consummate history and the eternal plan of God by rapturing His bride – the Church (i.e. the body of those who have believed in (and) received him and are serving him as faithful disciples).

Continuing, the document states, "A genuine saving relationship with Jesus Christ will result in a life of holiness and obedience." These declarations contain a number of key beliefs about Jesus held by the church: his eternity and incarnation, his life and death, his resurrection and ascension, and his second coming. He is the 'eternal second person of the trinity', 'the only mediator', the intercessor 'for his own', and the one who will consummate history at his return. In addition, it is declared that the outcome of a 'saving relationship with Christ' is 'a life of holiness and obedience'.

Nonetheless, as alluded to earlier, a major discovery during an eight-month period of observation of the three churches in their worship sessions, prayer meetings and bible studies is that most of their Christology is conveyed through songs. Hence, the beliefs contained in the preceding official statement are complemented by the Christological themes and imageries contained in a number of their songs. The basis of selection of the following songs for analysis is their frequency of use in the three churches, each of which was sung at least three times in various settings. First, Jesus is described as the way, the truth and the life, which is based on John 14:6:

> Jesus the way, the truth and the life
> Whosoever cometh to him
> Shall never die

Not only that, he is declared to be alive forever more:

> He's alive; Amen (2x)
> Jesus is alive
> Forever, he's alive; Amen

In addition, it is declared that all power belongs to him as the following song shows:

All power belongs to Jesus
All power belongs to God
All power belongs to Him
All power belongs to God

This theme of Jesus' power is further carried by a song that was found to be popular among the children in Faith Baptist. It is a two-line chorus that says,

Jesus is a bulldozer, Amen
He's a bulldozer

Because he has all power, he is able to do wonders, not just in a generic sense, but for individual Christians. This is conveyed in the song,

*Na so so wonda Jesus dey do – O!*
*Na so so wonda Jesus dey do!*
*Oh he has done it for me (3x)*
*Na so so wonda Jesus dey do!*

However, the most popular Christological songs in the three churches are the ones that extol and declare him or his name to be higher than all others. This could be deduced from the following two songs:

The Most Excellent Name is Jesus
Shout halleluiah, Amen
The most excellent name is Jesus
Shout halleluiah, Amen

His name is higher, above other names,
His name is Jesus, his name is Lord,
His name is higher, above other name,
His name is Jesus, His name is Lord.

The position of Jesus vis-à-vis the believer and every other person or authority is not lost out on them. Hence they declare him to be *Oga* and *Oga-Kpatakpata* through two significant songs:

*Oga Jesus you dey sweet me for belle,*
*You do good for me*
*You dey make me wonder,*
*Oga Jesus you dey sweet me for belle,*
*How I for do–oh*
*Jesus na wa yah - - oh*

*Jesus, na you biko,*
*Oga-kpatakpata,*
*na you biko,*
*Oga-kpatakpata*

A fuller explanation of '*Oga*' and '*Oga-Kpatakpata*' is given below; but suffice it to state here that they refer to the big-man, or the boss, and the over-all boss or chief executive, respectively, in any relationship or organisation, whether social, corporate or political.

In addition, Jesus is addressed as 'the Great One', the 'immortal God' and 'the immortal king', as the following Yoruba song attests:

| | |
|---|---|
| *Kabiesi – Oh Hossana O ---* | Great one Oh – hosanna O --- |
| *Eyin Jesu Kristi* | Praise Jesus Christ |
| *Oba a yiku,* | The immortal God |
| *Egbe ga – eyi Jesu O* | Praise Jesus Christ |
| *Eyin Jesus Kristi oba ai – ku* | The immortal King |

The last song to be mentioned is in Kalabari and it depicts Jesus as 'Saviour':

| | |
|---|---|
| *Jezos ani-iburoma bo* | Jesus is my Saviour |
| *ajike biya* | I will not be moved |
| *ori werisobiaajike biya* | He will not leave me alone |
| *toru akaka dibisime-semebra* | like the tree planted by the river side |
| *ajike biya* | I will not be moved |

It is pertinent at this point to locate this discourse within the context of African Christological scholarship. In academic African Christian Theology a variety of answers has been given to the question Jesus posed his disciples, "Who do you say I am?" The question has generated so many answers that Charles Nyamiti has rightly asserted that Christology is the most developed area of African Christian Theology (Nyamiti 1994, 70), while John Baur sees it as the central theme in that theology (Baur 1994, 304). This development signals the replacement of the Christological *crisis* that characterised the first phase of modern African theology (1950s to 1980s) with a new found *confidence* in the second phase (1980s to the present) that built on the foundations of the first phase (Stinton 2006, 6-13).

James Okoye identifies four major approaches adopted by African theologians in their quest to contextualise Jesus and his message to the Continent. They are *the comparative*, which moves from the Bible to African culture or vice versa; *the systematic*, which seeks a methodical and logical explanation that incorporates the nexus between the different mysteries of Christ and the church; and *the liberationist*, which is "not as interested in a Christ for worship and spirituality as in a Christ who enables change and transformation" and on the earthly Jesus and his unfeigned devotion to the cause of the oppressed as the cause of God. The last but not the least approach is *the community-based*, which considers the community's actual experience of Christ and seeks to distinguish between the imageries of Jesus which inform the people's real Christian living and the parts of the Bible which speak more personally to them (Okoye 1997, 9-17). Also, John Parratt identifies the following four approaches: (1) the "functional," or 'existential', approach that emphasises Jesus' deeds in relation to the individual believer; (2) the approach that sees his significance in terms of African charismatic leaders; (3) the approach that adopts the offices or titles used in traditional African societies, such as 'king' or 'chief'; and (4) feminist approach in which "Jesus is seen as a symbol for all humanity, female as well as male" (Parratt 1995, 81-91).

The outcome of these various approaches is a plethora of imageries that have emerged, such as *chief* and *okyeame*, or the "linguist," developed by John Pobee; *ancestor*, by V. Mulago, J. Penoukou, and especially C. Nyamiti (Christ as *brother-ancestor*), and Bénézet Bujo (Christ as *proto-ancestor*); *liberator*, by Jean-arc Éla, Eboussi-Boulaga, and Laurenti Magesa; *ruler*, by Francois Kabasélé; and *guest*, by Enyi Ben Udoh (see Stinton 206, 49-218). Each of these models is an attempt by their proponents to develop an authentic description of who Jesus is to the African.

However, some African theologians prefer a more eclectic approach, claiming that although such Christological categorisations have their merits, they are oppressive and reductionist (cf. Sawyerr 1968, 72-73). Also, they tend to follow the top-down pattern in which academic theology attempts to *prescribe* Christological categories for African believers. Another criticism against some of these Christological models is that they are merely academic and end in the classroom as a number of them have not been tested out at the grassroots level.

In the case of the data from the case studies, there are a number of Christological imageries that emerge. Such imageries derive from their daily experiences within the context of the Niger Delta and its peculiar challenges. They arise as the churches seek to find answer to the question of Jesus' identity within the context of conflict, since "genuine Christological reflection cannot be separated from Africa's socio-political, religio-cultural and economic contexts" (Akinade 1995, 184). In First Baptist he is affirmed as the Son of God, Saviour, Love and the Master of

Peace who stands for truth and justice and the downtrodden; in Faith Baptist, in addition to the imageries of The Great One ( or *Kabiyesi*), King (or *Oba*), Lord, and Saviour, two unique descriptions are highlighted, namely, *Oga-kpatakpata* and bulldozer; while in Hilltop Jesus is presented as the Eternal Second Person of the Trinity, Lord, Conqueror, Saviour, Mediator, and Intercessor.

At least three major facts emerge from the above analysis: first, the imageries of Jesus could be subdivided into two groups, the substantive and the functional. The substantive imageries, which include Son of God, King, Lord, and Saviour, derive from and are influenced by the NT; while the functional imageries are drawn from their daily experiences, especially within the context of oil and conflict, and include Conqueror, Love and Master of Peace, *Oga/Oga-kpatakpata*, and bulldozer.

Second, the humanity of Jesus predominates in their notion as they seek to make meaning out of their belief in Christ to their current situation. This is the same case with African theologians as noted by Parratt: "The humanity (of Jesus) has received a good deal of attention from African theologians, often with the stress on Jesus as the one who participates fully in the human community" (Parratt 1995, 82). Thus, he is seen as a people-centred Jesus who stands for the down-trodden. Third, it is obvious that their Christological imageries are influenced by three main sources: Scripture, experience, and the African worldview. However, because other imageries of Jesus mentioned in the case studies are generally known and for reasons of space, the Christological models of *Oga/Oga-kpatakpata* and Bulldozer are taken up further.

### *5.4.1 Jesus as Oga-Kpatakpata*

In Nigerian common parlance, *oga* refers to 'the big man', 'boss', 'owner', or 'superior', depending on the context. Superiority may be in terms of position, rank and/or status. Hence, an *oga* may be male or female. When used in reference to a female, it is prefixed to the word 'madam' as *oga madam*. Sometimes, married women use the expression for their husbands as a mark of deference. In a culture where calling an older person or a superior by the first name is abhorred, *oga* is used in informal and social settings as a substitute to the name of the person being addressed. Therefore, underneath its usage are informality, respect, and adherence to conversational protocol.

However, its usage in daily interactions also carries some negative connotations as it entails a subservient relationship between the *oga* and his/her subordinates and represents 'the big man syndrome'. Some writers have described this syndrome in various ways. For example, while Eberlein uses the expressions "Godfatherism/Godmotherism" to portray it (Eberlein 2006, 517), Haynes calls it "the big man, small boy syndrome" (Haynes 1996, 9; cf. Gifford 2004, 175). In this syndrome, the *oga* is served by all but serves none; he/she has access to and control of the lion-share of the collective resources of the people from which he/she 'dashes' small portions as he/she wills, and he/she is accountable to none except him/herself and his/her peers, namely, fellow *ogas*. As a consequence, several *ogas* are held in disdain by their people and such disdain is usually suppressed in their presence but openly expressed in their absence.

Not only is Jesus portrayed as *Oga*, he is described as *Oga-kpatakpata*. Whereas *oga* is commonly used to refer to any person in a position of authority, from the home to the Government or the corporation, *oga-kpatakpata* is reserved for the very few: the over-all boss, the Group CEO, the person who has the final say in any organisation or institution. The *oga-kpatakpata* is the *oga* of *ogas*, the one to whom all other *ogas* report to. Therefore the imagery carries the idea of a top-down hierarchy in which each succeeding level serves as a channel of communication until it reaches the *oga-kpatakpata*.

However, in the circumstance that a junior officer feels mistreated by his immediate *oga*, he could then appeal his case to the *oga-kpatakpata* from whom he expects to receive fairness and justice. Theologically, it corresponds to the biblical notion of 'King of kings' and 'Lord of lords', and is related to the Baptist doctrine of the absolute lordship of Christ. George Truett describes this doctrine as "the nerve center of all their Christian life, the bedrock of all their church polity, the sheet anchor of all their hopes, the climax and crown of all their rejoicing" (Truett 2008, 147). From the perspective of African contextual Christology, it falls under Okoye's *systematic* and *liberationist* approaches and to Parratt's existential and title approaches and specifically relates to Pobee's *chief* idea, to Ela's *liberator* notion, and to Kabasele's *ruler* Christology. Other Christological models from the case studies that can be subsumed under this imagery since it captures their core tenets are *Kabiyesi*, *Oba*, Lord, Conqueror, and Saviour.

Behind this imagery are certain emphases in the thinking of the church members. First, it carries the idea of cordiality and informality that exists between the believer and his/her Master, Jesus. Second, it underscores the regard which the believer accords him so as not to obliterate the fact that he is the one in charge. And, third, it emphasises the strongly held notion that Jesus is the final arbiter in matters of life, including the Niger Delta conflict. This is further underscored by the two songs, 'The Most

Excellent Name is Jesus' and 'His Name is Higher'. He is the one to whom they run for justice, provision, protection and peace. But, unlike other *oga-kpatakpata* that are detached from those below them, Jesus is intimately connected to his people and meets their needs to such an extent that they are amazed at him; thus they sing, *"you dey sweet me for belle, you do good for me, you dey make me wonder!"*

### 5.4.2 Jesus as Bulldozer

The second major representation of Jesus describes him as a bulldozer. It is expressed mainly through a song which is especially popular among the little children in Faith Baptist. The song simply says, *"Jesus is a bulldozer, Amen, He's a bulldozer, Amen!"* The bulldozer is a heavy earth-moving machine used in the construction industry to clear forests, excavate refuse, and pull down structures. Nothing usually withstands its force in a construction site. Thus, although it might be destructive in its activity, especially against the environment, it ultimately produces positive results as it prepares the ground for other works. This imagery, though imperfect, represents in the mind of the people three basic things: power, clearing and destruction of debris and obstacles, and creation of order out of chaos and uselessness. Therefore, Jesus represents for them power over all forces, the ability to destroy all evil powers attempting to impede their wellbeing, and the one who can create hope out of hopelessness. Nigerian Catholic theologian Agbonkhianmeghe Orobator also notes the presence of this model among a group of prison inmates in Benin City, Nigeria for whom "only a *Jesu Kristi* endowed with the power and force of a bulldozer would do" (Orobator 2008, 77-78).

### 5.4.3 Political Implications of Jesus as Oga-Kpatakpata and Bulldozer

A pertinent issue that arises in the light of the foregoing discussion is the political implications of the belief in Jesus as *Oga-kpatakpata* and Bulldozer for Nigeria's political economy of oil. This will be conflated with the notion of Jesus as Conqueror, Saviour and Love and Master of Peace who stands for truth and justice. To reiterate, these imageries aggregate to four themes: power, domain, delegated authority, and relationship. Power, domain and delegation of authority are particularly seen in the conceptions of Conqueror and *Oga-kpatakpata*. He is seen as the conqueror of Satan and his cohorts and indeed any other authority or power that is opposed to the purpose, plan and principles of God. Not only that, he is the *Oga* of all *ogas* and his "*ogaship*" covers all domains of existence, whether seen or unseen, both the spiritual and the political. Therefore all other *ogas* are answerable to him and will eventually do so.

Moreover, as every *oga-kpatakpata* normally has those that represent and act for him/her by way of delegated authority, Jesus as *Oga-kpatakpata* delegates authority to his church in a diffused system of authority flow. The authority flow could be democratic, as in the two Baptist churches in this study, or hierarchical, as observed in Hilltop. Among the Baptists, authority resides in the congregation because of the belief that "the church is to be governed not by an order of priests, nor through higher or central courts, but through the voice of the Holy Spirit in the hearts of the members in each local assembly" and "the equality of status of every church member and the recognition of the diversity of gifts," while gladly accepting the guidance of its chosen leaders (Kevan 1984, 123). Notwithstanding the hierarchical structure of Hilltop, a close observation of its operations shows that it is egalitarian in ministry and relationship because of its belief in the equality of all believers in Christ and the free distribution of spiritual gifts among the members.

Furthermore, from a theological perspective, Jesus' role as *Oga-kpatakpata* extends to the realms of history and culture. That is, he is regarded by members of these churches as the *Oga* of history and culture. This could be seen in their theologies of prayer and conversion. This is indicated in the former through the two peculiar understandings of prayer as declaration and as prophecy. By praying in either form, they believe they are standing in the name of Christ, as his ambassadors, to bring to bear his will in particular circumstances via the instrument of the spoken word (see the next chapter for the analysis of their ecclesiological imageries). For example, when instructed to prophesy or declare peace for Port Harcourt, members of Faith Baptist in a prayer meeting individually and vociferously declared, "In the name of Jesus, I prophesy peace for Port Harcourt"; while in Hilltop, on one occasion, a prayer leader declared, "In Jesus' name (the congregation shouted Amen!), there shall be no miscarriage." Behind this mode of thinking is the belief that Jesus has power over events in space and time and to speak in his name is to activate that power. The question of whether such prophecies or declarations are in accordance with the will of God, or whether they actually come to pass in real terms is not an issue for them. Their theology of conversion also shows a modicum of belief in the supremacy of Jesus over history. This is with particular reference to the *ontological*, *political* and *existential* aspects of that theology as will be outlined in the next chapter.

Perhaps the most problematic among these power-domain imageries is that of bulldozer for the bulldozer as a heavy duty machine is used to destroy trees, houses, and other structures in readiness for some constructive work. Nonetheless, it connotes for them Jesus' ability to destroy all negative forces and thereby prepare the way for the good and positive to prevail. In other words, Jesus first destroys all obstructions to fulfilling his set purpose and out of that destruction produces something that is constructive and good. Nothing, or nobody, can withstand his power or thwart his purpose since 'All power belongs to Him'. It therefore complements the imageries of Conqueror and *Oga-kpatakpata*.

Also, the imagery of Jesus as *Oga-kpatakpata* is complemented by those of Love and Master of Peace who loves truth and justice, which emphasise that he is different in his approach from human *ogas*, or 'The Big Man'. He exercises his power within his domain, a domain that is universal, in love, peace, truth and justice. That is, Jesus is seen as the embodiment of the virtues of truth and peace, and holds in perfect balance power and love.

A major political implication of the above interpretation of the *Oga-kpatakpata* Christology, and which could serve as a key theological contribution of this study to Nigeria's overall political economy, is a redefinition and right application of power. If indeed Jesus is the *oga* of *ogas* and yet exercises his power and authority in love, truth and justice, then Nigerians, generally, and Christians, in particular, ought to borrow a leaf from this, such that it is understood that the true test of power is its effectiveness and service, rather than control and domination (see Gifford 2001, 17). This theme is taken up further in Chapter Eight in the proposal of a political theology for Nigeria's political economy of oil and conflict.

### *5.4.4 Jesus and Violence*

A major topic that attracted extensive discussion during the focus group sessions, especially in First Baptist and Faith Baptist, was on their members' perception of Jesus regarding violence, particularly within their immediate socio-political context. A number of members in both congregations espoused the view that Jesus stands for peace. The following comments are representative of this. The first is by a mother in First Baptist who comments as follows:

> To me Jesus is the Master of peace. When he was at the Sea of Galilee he maintained peace. In his way of doing things, he wept with those who were crying and even laughed with those who were enjoying. So Jesus is the Master of peace who settles every matter in a just way. Since he

> preaches love and peace Jesus would not want anybody to kill another; rather he would encourage us to be our brother's keeper. Therefore ... I know very well that he supports peace, love and justice, since he is the Son of God.

An elderly female member of the church tows the same line of thought by declaring thus:

> I believe Jesus is the Son of God who preaches love and peace; he would never want anybody to kill another and he would encourage us to be our brother's keeper. On the basis of what I know very well, I'm sure that he stands for peace, love and justice. These are the qualities that stand him out.

Yet another female member who is a petty trader adds, "Jesus is someone who maintains peace to the end. When he was at the river he said, 'let there be peace' and there was peace."

Also a middle-aged house wife in the church cites Jesus' reaction to Peter's action at the Garden of Gethsemane when the mob came to arrest him to substantiate her position that Jesus is a person of non-violence. Her words:

> When Peter cut off the ear of the servant of the Chief Priest, Jesus asked him to put back his sword into its sheath, that if he wanted to fight, he would ask his Father to send angels to fight for him; so he should not use his sword. Therefore, in the fighting going on in our place, youths should drop their guns because Jesus does not like violence.

Conversely, there are also those in First Baptist who believe that Jesus supports violence as he is alleged to have become angry at certain points as recorded in some New Testament accounts such that he was even violent. Thus, in response to the immediate preceding contribution above, a young man in the church says,

> In my own (view), when Jesus saw Peter cut off the ear of Malchus, he asked him to put back his sword. But I want to say that the real primary aim why Jesus came to this earth had not been done. Jesus came to deliver the children of Israel from bondage and that's why he came to this earth. But that has not been done and that's why till now most of the Israelites have not accepted him as their Lord and Saviour because they felt that his coming would bring total liberation to them. Now to apply it to our Niger Delta case ... how can we the owners of the land sit, and see others come to eat our food, and when they eat, instead of them to keep some for us, they take all and we the owners of the food go to beg. So the Niger Delta people can carry arms so as to fight for what rightfully belongs to them.

Another male member, who is a fisherman, declares as follows:

> The action that Jesus took when he went to the temple is the reason why the Niger Delta people are taking action too. When Jesus went to the Temple, which was meant for worship, he saw people exchanging money and trading. When he came in, he was very angry, that that was his Father's house; they were not supposed to do those things there. He was very angry and he drove them out of the Temple. So this is what the Niger Delta is doing now. They are trying to fight for themselves.

A middle-aged male member agrees with him by saying: "We could remember some instances in the Bible when he grew annoyed and used whip to flog people in the church. So, with this, I can say that Jesus somehow supports the use of some violence and I know that if he were here today, he would support the south." However, a middle-aged woman retorted: "No, Jesus' action in the temple does not mean he supports violence. He only pursued the money changers and flogged the animals. He is the prince of peace and so cannot use violence on his enemies."

When a poll of the members of First Baptist present at the focus group session was conducted, about eighty percent (87) agreed to the Jesus-for-non-violence view, while the rest (23) subscribed to the Jesus-for-violence position. The pattern that emerged from the poll was that virtually all those in support of the latter view were males, while those that agreed to the former were females. A further dynamic that was noticed was that the females were on average older than the males.

In order to further tease out this discussion, a hypothetical question was then posed, namely, "From your own perspective and from your understanding of the portrayals of Jesus in the New Testament, if he were physically present in the Niger Delta, what do you think he would have done or said regarding the crisis?" The question was aimed at determining whether or not they had any view on the possible political implications of Jesus' social ethics for the conflict, especially in the light of their Christological descriptions.

Once again, the answers were diverse. First, an elderly male member of Faith Baptist who is a retiree believes that Jesus would not have endorsed the killings taking place in the region; rather, he would have insisted on the practice of true federalism in the country. Hence he says:

> If Christ were here today, he would have said, 'don't kill', and he would have also said, 'you should practise federalism in your presidential system of government, as it was in the days of old when the derivation principle was the mainstay of this country, and it brought peace. When groundnut was being produced the sharing was 50% for the region and 50% for the

> nation; then oil should also be the same so that this people will not be aggrieved.' I think that would have solved the whole problem now; the sentiment wouldn't have arisen; sectionalism wouldn't have been there; Christ would have done that. The only way which you can quench the thirst of these people who are agitating is to go back to that thing which had been bringing peace to this country.

A youth in the church who is a lawyer focuses his response on what he considers to be the appropriate thing Jesus would have done, namely, the amendment of certain laws that regulate the oil industry in Nigeria. Therefore he says:

> I believe that if Christ were to be here, he would have gone to the root of the matter to make fundamental changes so that there would be lasting peace. We should remember that this issue of the Niger Delta did not start today. The fears of the Niger Delta did not start today… if Christ were to be here today, what he would do, first of all, is to look for the Petroleum Act and remove that section that says that everything belongs to the Federal Government. Christ would look for the Land Use Act that says the use of land that belongs to your village is to be determined by the Governor. He would say 'No', I'm the one that located you here; you should have a right on how you use it'. So he would amend that section. Christ would make fundamental changes and install justice because until we do that, the establishment of NDDC, OMPADEC, and the rest of them would not solve the problem.

However, a female member in the church, who is a medical doctor, argues that Jesus' primary focus would be the religious people and that he would not condemn anybody but would rather speak the truth to every person concerned, 'boldly, 'lovingly' and 'impartially'. The following are her words:

> Let's look at the Gospel, we know that God has a plan that is unfolding and it depends on what God has. The Lord would always move with God. First of all, if we look at how Jesus treated the people of Israel, he first blamed the religious people. He said to them, 'woe unto you Scribes and Pharisees, you would traverse hell and earth to make a proselyte a worse candidate of hell than yourself'. They made people worse; in fact he would be preaching the Gospel. The only answer is because people don't understand, including church people like us. If we really listen to ourselves, we have not understood the heart of the Gospel. Because the Lord would not condemn anybody but he would condemn everybody's action and everybody would know that they have sinned. First the church people: the Bible says 'if my people which are called by my name will humble themselves and seek my face and turn away from their wickedness and pray, I will hear from heaven and I will heal their land.' … So if the

> Lord were here he would first condemn the action of religious people. In fact, the crux of the matter is with us. The Lord would go to the people involved and, without fear, but in love, would tell them the truth boldly, lovingly and impartially, unlike most of us who would be afraid to do so. He would also talk to the poor people.

The last part of the submission above is also expressed by an elderly mother in First Baptist who says, "Jesus would maintain peace because he would make the peace truthfully, that is, he would not side with any party; rather he would appropriately apportion to each group what is their due and he has said we should leave vengeance to him"; while a youth in Hilltop says, "I know two things about Jesus: truth and reality. So for the problem in the Niger Delta, Christ would begin with the truth and would stand upon reality; he would not fake and he would not deceive. In order to satisfy all, he would maintain peace based on truth and reality."

Also, a lady in Faith Baptist, who is a civil servant, believes that Jesus would have brought peace between the government and the people of the region through dialogue, since he 'does not have interest in violence'. Thus she says:

> I would say that if it were Christ, if Christ should come into this matter, he would bring peace between the Government and the Niger Delta. How? Because Jesus Christ does not have interest in violence, and that is why on the day he was arrested, when Peter brought forth his sword to cut off Marcus ear, he wasn't happy about it. So he wouldn't support violence, rather he would go about dialogue.

Since the questions of justice and injustice are at the heart of the crisis, a number of members of the three churches are of the view that Jesus, being the Son of God, would have established justice and eradicated injustice. To an elderly male member of Faith Baptist the way he would have done this is by distributing the oil resource across the nation so that all the regions would have it. So he comments thus:

> If Jesus were here today, I'm sure he would have toed the line of justice. Why it has been very difficult to establish justice in the circumstance is because we have been lucky or unlucky to have the oil and I'm sure Jesus would have decided to do something, because he is God, to spread the oil around the whole of this country. There would have been oil established; Jesus would have placed oil at the Chad Basin; Jesus would have also placed oil at the Benue Trough and other sections of the country where there is no oil. If oil were found in all these other places, we would not have any problem. That's what Jesus Christ would have done.

An unemployed youth in Hilltop comments as follows:

> If Christ were here, he would address the injustice in the Niger Delta. One, there is an altar speaking against the Niger Delta, it's an evil altar and that altar has to be destroyed. That is the root cause of the problem which has to be destroyed. So Jesus Christ being God would not support injustice but would destroy the evil altar.

Similarly, an illiterate middle-aged mother in First Baptist, speaking through an interpreter, says:

> If Jesus were on here on earth and see what the Niger Delta is going through, he would give justice. The truth of the matter is that Niger Delta people are really suffering… The north is pressing the south. So if the north gives justice to the south then there would be peace.

On the basis of the above answer by a member of his church, the Pastor of First Baptist directly asked the church in Kalabari in order to fully ascertain that they understood the discussion, "If Jesus were here on earth, would he have supported the violence going on in our region?" The chorus answer from the entire members present at the focus group session (109) was 'No!' The answer then raised another problem for this researcher because among those that said 'No' were persons who had earlier argued in favour of a violence-prone, violence-supporting Jesus. One possible explanation for this apparent discrepancy could be that they might have gained further insight into the issue in the course of the discussion, which led them to a change of mind. Or, perhaps, to them, although Jesus might endorse the use of violence under certain circumstances, it does not apply to the Niger Delta case.

## 5.5 The Holy Spirit

An interesting discovery that was made in the course of field research was that the Holy Spirit is the least mentioned member of the Trinity in the three congregations in their oral expressions within the context of conflict. For instance, whereas the category of God featured about 112 times in the worship sessions, focus group discussions, personal conversations, songs and prayers, Jesus Christ about 90 times, the Holy Spirit was directly mentioned a mere 30 times. But the reverse was the case regarding their written documents. For example, whereas the Baptists, through their official Statements of Faith, devote six lines to their affirmations on God and about a page to their beliefs about Jesus, about three pages are allotted to their basic teachings on the Holy Spirit and his fruit and gifts. The first part of the statement declares,

> God manifests Himself as Holy Spirit. God as Holy Spirit inspired the writers of both the Old and New Testaments. The Holy Spirit illumines the mind of man to enable him to understand spiritual truth. He convicts of sin, of righteousness and of judgement. He calls men to repent of sin and come to the Saviour. He effects regeneration (the new birth). He cultivates Christian character and bestows the spiritual gifts by which believers perform their service in the church. The presence of the Holy Spirit in the life of the believer is the assurance that God will bring the believer into the fullness of the stature of Christ. The Holy Spirit comforts the believer in times of trouble and enlightens and empowers the believer for worship, evangelism and service.

Hilltop, on its part, through its own Statement of Faith, declares concerning the Holy Spirit,

> The Holy Spirit was sent into the world by the Father and the Son to apply to mankind the saving work of Christ. He enlightens the minds of sinners, awakening them to recognise their need of a Saviour and regenerates them. He, the Holy Spirit, permanently indwells every believer, becoming the source of assurance of salvation, and uniquely endows each one with the gifts for the building up of the body. Every believer needs to recognise and must embrace the pentecostal empowerment of the baptism with the Holy Spirit with the initial evidence of speaking in other tongues as the Spirit gives utterance.

The above two sets of statements on the Holy Spirit share some similarities and slight differences. Both of them affirm the deity of the Holy Spirit and declare certain functions he carries out in the lives of individuals, such as illumination of the mind, bestowal of spiritual gifts, assurance of faith, and comfort in times of trouble. However, while the Baptist declaration mentions his role in the formulation of the Christian canon, the conviction of sinners, and empowerment for worship, evangelism and service, that of Hilltop is silent on them. Conversely, whereas Hilltop states the imperative nature of 'baptism with the Holy Spirit' with speaking in other tongues as its initial evidence, the Baptist document does not mention it. This could be illustrated with the following Table:

**Table 1 Affirmations on the Holy Spirit in the Case Studies**

| Affirmations on the Holy Spirit | Baptists | Hilltop |
|---|---|---|
| 1. Affirmation of his deity | Yes | Yes |
| 2. Role in relation to the Scriptures | | |
| – inspiration of both Testaments | Yes | No |
| 3. Roles in relation to individuals – | | |
| a. illumination of the mind | Yes | Yes |
| b. conviction of sin, | | |
| righteousness and judgement | Yes | No |
| c. cultivation of Christian | Yes | No |
| character | Yes | No |
| d. bestowal of spiritual gifts | Yes | Yes |
| e. assurance of faith | Yes | Yes |
| f. comforter in times of trouble | | |
| g. empowerment for worship, | Yes | No |
| evangelism and service | | |
| h. empowerment with 'baptism | No | Yes |
| with the Holy Spirit' | | |

The preceding statements are complemented by a number of prayers to and declarations about the Holy Spirit made through songs. Of interest were the settings, and especially the immediate context of such settings, wherein the songs were rendered. Some of the immediate settings are during the initial stages of worship; during prayer sessions, as part of the background music; just at the beginning of the sermon; and/or during the time of response by the congregation to the sermon generally known as 'alter call'. Some of the songs include the following:

Holy Spirit, do your own thing now
Do your own thing now
Holy Spirit, do your own thing now
Do your own thing now

And

Holy Ghost, do it again
Do it again, in my life
Open my eyes, to see Jesus
Seated on the throne

The above two songs were sung in Faith Baptist on three different occasions during the period of research, namely, on Sunday July 8, 2007, and on two Wednesdays during their 'Breakthrough Hour' and Last Wednesday worship of July 11 and 25, 2007, respectively. They were also sung in Hilltop on Wednesday September 6, 2006 and on Sunday August 5, 2007. The context when they were sung in Faith Baptist on July 8, 2007 and July 25, 2007 was during the period of intercession for the region, for peace to be restored, for the government at all levels and local chiefs, for them to be God-fearing. Both songs are prayers addressed directly to the Holy Spirit in which the worshippers are asking that he do 'his own thing' and to do 'it' again by opening their eyes to see Jesus seated on the throne. This elicited in the researcher the desire to fathom the possible connection between the songs and their timing in their mindset, vis-à-vis the relevance of the Holy Spirit to the conflict in the region. Some deductions were drawn on this after much reflection.

First, that they see the Holy Spirit as deserving of direct address since he shares the same 'nature, essence and being' with other members of the Trinity. Second, it shows that to them, the Holy Spirit is the one that can do his 'own thing' in the region. 'His own thing' in this context refers to peace. Thus, although they are praying to God for peace in words, they are saying the same prayers in songs to the Holy Spirit. In this context, therefore, God and the Holy Spirit are one and the same. The third deduction is that, in line with their belief that the Holy Spirit enlightens the mind, they are asking him to enable them see Jesus as the ultimate ruler despite the conflict caused by the misrule of earthly leaders in the country, beginning with local chiefs.

The context they were sung in Hilltop on Wednesday September 6, 2006 was immediately after the message by the Senior Pastor when he was calling people out to the rostrum for various decisions, such as salvation, rededication and sanctification, and breakthrough in their lives. In this setting, the Holy Spirit's 'own thing' refers to convicting people for regeneration, commitment to righteous living, and power for success and victory.

The same thing applies to the next song:

Anointing, fall on me (2x)
Like the fire of the Holy Ghost,
Fall on me,
Anointing, fall on me

This song was particularly sung with passion and intensity by the congregation at Hilltop during its 'Anointing Service' on Sunday August

5, 2007. It was immediately following by a seventy five-minute sermon by the Senior Pastor as they started the actual anointing of worshippers with 'the anointing oil', which is made of olive oil. As described in chapter four, just before the ushers started distributing the oil among the congregants, the Pastor declared, "God is pouring out liquid fire through this oil. No demon can withstand it." Continuing, he added,

> The oil symbolises the Holy Spirit; and as the Bible says, it is only the anointing that can break the yoke. Our faith is not in the oil but in the one it represents. The Holy Spirit shall empower you for success, for healing, for victory, for business breakthrough. No power can withstand you; childlessness and miscarriage cannot hold you back; HIV/AIDS cannot stop you; you are unstoppable; you can make it, on account of the anointing. This week, go and submit that application; you'll get the job.

In an interaction with one of the pastors of the church after the service, when he was asked to expatiate on what he considered to be the role of the anointing in the Niger Delta conflict as described by the Senior Pastor, he said,

> Indeed it is the anointing that can break the yoke of militancy in the region by delivering the youths from the bondage of cultism, drug addiction, kidnapping and armed robbery. Without it, permanent change cannot take place in the lives of these young people. Even if they renounce militancy today, without the help of the Holy Spirit, they might go back to their former life. So, we need the anointing of the Holy Spirit even in this matter.

Thus, anointing, which is believed to come from the Holy Spirit, is basically for power over demons and over any success-negating factor in life, such as sickness, barrenness, and joblessness, as well as for the transformation of individual lives.

## 5.6 Summary

This chapter has shown that the major sources of the theologies of the three congregations within the context of oil and conflict are the Scriptures, their denominational affirmations, the African world view, and their individual and collective experiences. These correspond roughly to the three basic elements of African theology identified by John Parratt, namely, the Bible and Christian tradition, African culture and religion, and the contemporary socio-political situation in the continent (Parratt 1995, 27).

It was discovered that the main recurring theological themes in the case studies within the context of the Niger Delta conflict are God, Jesus Christ, the Holy Spirit, Prayer, the Church, and Conversion. The first four were treated in this chapter, while the remaining two will be discussed in the next. Moreover, their theologies are conveyed mainly through songs (33%), prayers (25%), and sermons and statements (42%). God is affirmed as *Tamuno*, *Baba*, *Papa*, *Ogologo Mma Agha* ('Long', or 'Mighty', Sword), as well as the God of Abraham, Isaac and Jacob, a Man-of-War, and the One in control. He is also seen as Punisher, Fighter, Avenger, Enabler, and the One who blesses and answers prayers. Four blocks of attributes of God were identified from the data, namely, absolute, authoritative, affable, and adversarial attributes. The belief that God is the One that answers prayers was so strong that prayer was found to be their single most predominant mode of response to the crisis. Hence, their theology of prayer was analysed.

On the category of Jesus Christ, he was conceived in various ways as *Oga Kpatakpata*, Bulldozer, *Kabiyesi* (Great One), *Oba a yiku* (Immortal God), Son of God, Saviour, Love and Master of Peace, Eternal Second Person of the Trinity, Mediator, Intercessor, and Conqueror. Due to their uniqueness, the models of *Oga Kpatakpata* and Bulldozer were further discussed, alongside their political implications for Nigeria's political economy of oil and conflict and their relationship with some of the leading Christologies in African Christian Theology.

Regarding the Holy Spirit, a paradox was discovered: while he is the least mentioned member of the Trinity in their oral theologies, he clearly predominates in their Statements of Faith in terms of the volume of space he occupies. Nevertheless, he is considered in a variety of ways as the One that empowers, anoints, and enlightens the mind. The next chapter will dwell on the category of the church.

# CHAPTER SIX

# THE CHURCH: MEANING, PURPOSE, AND SOCIAL ROLE

## 6.1 Introduction

This chapter focuses on one of the major objectives of this book, namely, the examination of the validity, or otherwise, of the claims by Paul Gifford that African evangelical Christianity lacks social responsibility, is anti-development, apolitical, and only interested in soul winning. Thus, the chapter will describe and analyse what the three congregations in the study are saying concerning the Church within the framework of oil and conflict; whether or not they have any sense of social responsibility; and what they are doing, if anything, within the context of the conflict. As earlier indicated in chapter four, the category of Church was selected based on its occurrence across the three churches.

The major sub-themes that emerged from the data under this category based on their frequency and distribution are its meaning and purpose, Christian social responsibility and political involvement, and conversion (evangelism and soul winning). The total appearance of the subcategory of the meaning and purpose of the church across the board is 136, distributed as follows: 39 in First Baptist, 35 in Faith Baptist, and 62 in Hilltop. Christian social responsibility featured ten times in First Baptist, twenty times in Faith Baptist, and fifteen times in Hilltop.

## 6.2 The Church

As in the case of the categories in the previous chapter, this section will begin with an analysis of the official statements of the congregations on the Church.

### 6.2.1 Official Statements on the Church

In its "Statement of Faith," Hilltop describes the Church as follows:

> The second part of union with Jesus Christ is that all believers become members of his body, the church. There is one true church universal, comprised of all who acknowledge Jesus Christ as Saviour and Lord. The scripture commands believers to gather together to devote themselves to worship, prayer, teaching of the word, observance of baptism, communion established by Jesus Christ, fellowship, service to the body through the development and use of gifts and talents, and outreach to the world.

The statement describes the Church as "the second part of the union with Jesus Christ" wherein all believers become members of his body. Thus, it uses the image of the body to describe the church, which it says has two aspects: the one, true universal church made up of all who acknowledge Jesus Christ as Saviour and Lord, and the local church made up of individual believers. According to this statement, the purposes of the church include worship, prayer, teaching, observance of baptism and communion, fellowship, service, and outreach.

The official Baptist declaration on the Church is as well contained in their Statements of Faith and Practice. It declares as follows:

> The church of Christ as seen in the New Testament is a body of baptized believers in Christ who have covenanted with one another to worship, fellowship and serve God through Christ. A local church may be defined as a body of baptized believers in Christ organized on democratic principles under Christ as head, meeting together for worship, fellowship, Bible Study, prayer, and to observe the ordinances of baptism and the Lord's Supper while carrying out Christ's commission to evangelize the world. In such a church members are equally responsible. Through the local church the believer expresses his life, gifts and talents. The local church is the one most mentioned in the New Testament although it also speaks of the church as the body of Christ which includes all of the redeemed of all ages. Every local church under the leadership of Christ also interdependently works with others of like faith and understanding in order to carry out Christ's commission to make disciples of all nations, teaching them to observe all things commanded by Him.

The above statements have close affinity with those of Hilltop. For example, just as in the case of the Hilltop statement, the Baptist statement uses the imagery of the body to describe the Church, acknowledges the two aspects of the Church – the local and the universal, and affirms the same purposes of the body of Christ, namely, worship, fellowship, teaching, prayer, observance of the ordinances, and evangelism expressed as carrying out Christ's commission. But the Baptist statement adds three other affirmations: the voluntary nature of the Church in its composition, its democratic ethos in polity, and interdependence with other churches with "like faith" for obeying Christ's commission. The document also explicitly states that Christ is the head of the Church. The above official declarations perhaps inform the statements of members of the three churches on the meaning and purpose of the Church.

### *6.2.2 Meaning of 'Church'*

Apparently drawing from their church's definitions of the Church, three members of First Baptist (a mother, a middle-aged man and a student) describe the Church, respectively, in the following short sentences: "The Church is a body of believers," "The Church refers to the people who have accepted Christ and worship him," and "The Church is the called out ones that have accepted Christ as their Lord and personal Saviour." In Faith Baptist, out of sixteen specific definitions of the Church offered by some members, nine refer to a group of people defined by their personal relationship with Christ; while seven refer to place. For example, an elderly woman who is a retired civil servant says, "A church is a congregation or collection of people who believe in the saving grace of our Lord Jesus Christ who died for us." Other descriptions are as follows: "The Church is a group of local baptised believers in Christ who voluntarily bind themselves together to carry out the work of Christ" (by a middle-aged mother); "The Church … is a collection of believers, or those that believe in Christ. Another aspect of the Church is that it is the representative of Christ here on earth" (by another middle-aged woman); and "The Church can be defined as a body of believers gathered together to worship God" (by a self-employed middle-aged man). While referring to Matt. 16, a male university student adds as follows:

> A church is made up of individuals. That is, when Jesus said, 'Thou art Peter, and upon you I will build the church and the gates of hell shall not prevail,' he was referring to a body of individuals. Also the church is the universal body of Christians as well as a group of baptized believers coming together for fellowship in a particular place. Also the structure where we gather to worship is the church.

The following descriptions were offered by members of Hilltop in a focus group discussion. The first is by a young university graduate: "The Bible says that Jesus told his disciples, 'I will build my church and the gates of hell will not prevail against it'. The Church is made up of individuals; we the people that are gathered here we are the people that make up the Church. The Church is not the building and the society sees us as the Church." The second contributor, a young lady in the church Choir, submits thus: "The Bible says that the Church is the body of Christ and if the body has only one head, it then means that everybody is being blended to one body. So if we operate exactly the way Christ wants us to operate, the Niger Delta would be a good place." Then a young man in his thirties adds:

> Let me just add to what my sister said that the Church is a body. As the body of Christ, God expects each one of us to carry out his or her responsibility. It is also written that we are ambassadors of Christ; we are here on earth to represent Christ. So as a church we have the responsibility of showing the world that we are acting upon God's word by representing Christ wherever we are and in whatever we do.

A group of twenty nine young boys between the ages of ten and twenty five in Faith Baptist offered descriptions of church that emphasise space/place. They include the following: "The Church is a place where we come and learn and understand what Jesus has said like in the Bible"; "The Church is where people gather together to seek the face of God and learn about God, may be learn from what has happened and take their corrections from other things that have happened as it is in the Bible"; "The Church simply means where we come to understand God more and do discoveries on his word and talk to him"; "a place where Christians come together to learn and teach about the ways of God"; "the Church is a place where Christians gather to worship God"; and "the Church is a place where we come to worship and relate with God and to also grow spiritually with him."

A unique description of the Church in the context of conflict was offered by the Pastor of First Baptist who describes the Church as "a city of refuge." He puts it as follows:

> At the peak of the battle of 2002 and 2004, our church served as a city of refuge for a large number of people as both members and non-members ran to the church for safety. Some of them had lost their houses and means of livelihood to the conflict and had no other place to go to. So they came to the church. Some of them stayed in the pastorium with my family, while others stayed inside the church building. My wife and I had to provide food for them out of our meagre resources. At a point during the fighting, even some militants, when they are being pursued by their opponents, ran into the church building for safety and their pursuers did not continue with the chase into the building.

(This researcher witnessed an example of this in 2002 while in the Church for a funeral service. A heavy gun battle had ensued between two rival gangs besides the church building while the funeral service was going on. At a point a member of one of the gangs ran into the auditorium as he was being pursued by his opponents. The pursuers did not continue the chase into the building.)

The above responses of the members on the meaning of the Church dovetail into two blocks of thoughts – those that stress the *being* of the Church and those that emphasise *space*. The ontological definitions use words that refer to groups of persons while the spatial descriptions put the stress on place. A pattern that was noticed in the course of the research was that most of the answers from the youth tilted more towards spatial understanding, while the elderly ones placed more emphasis on ontological expressions. A further analysis of these definitions and/or descriptions of the Church yielded the following data:

- Body imagery – body of believers, body of Christ: 10
- Imagery that emphasise community or collection of people, such as exemplary community, group of people, congregation of baptized believers: 9
- A place or structure for a variety of purposes: 8
- Ambassador, or representative, of Christ: 3
- City of Refuge: 1

The fact that the imagery of the body for the church is the single most occurring model in the three churches is not surprising, especially among members of the two Baptist churches. This may not be unconnected to its historical usage among Baptists. For example, Article 3 of the Schleitheim Confession of 1527, in which Anabaptist leaders delineated seven articles of faith, describes the Church as the "one body of Christ which is the church of God and whose head is Christ" (Slatter 1999, 43; cf. Yoder 1977, 11). Also, several Baptist Statements of Faith use it. For example, the Southern Baptist Convention uses the description with slight variations in its 1963 and 2000 versions of *The Baptist Faith and Message*, while the Nigerian Baptist Convention describes the Church as "a body of baptized believers in Christ who have covenanted with one another to worship, fellowship and serve God through Christ" in its own *Statements of Faith and Practice*.

Furthermore, the above responses bring out some salient points in the members' understanding of the meaning of the Church. First, five emphases predominate, namely, *embodiment, community*, *space* and *functionality*;

second, the means to becoming a member of the Church is a personal commitment to Jesus; third, there is the belief that the Church does not live only for itself but is supposed to represent Christ on earth; fourth, membership in the Church is voluntary and not coerced and cannot be obtained by proxy; and fifth, this voluntary membership carries with it an obligation, namely, to carry out the "work of Christ."

In the understanding of members of the three congregations, a local church refers *primarily* to the people that make it up rather than to the building. The individuals who constitute it collectively coalesce into one body, the body of Christ, and thereby become a separate community. This links the emphasis on embodiment to that of community. In their estimation the Church as a community is the *spatio-temporal*, corporeal and tangible representation of Christ on earth. As such it is to serve as his ambassador. Functionally speaking, the Church is meant to carry out its ambassadorial responsibilities to the world by showing it the way it should go, a way that is based on God's word.

A juxtaposition of the above points would produce the following definition of the Church from the oral theology of members of the three churches: The Church is an embodied community of baptised believers in Christ who voluntarily come together in a particular place to serve as his ambassadors and representatives to the world with a view to carrying out the 'work of Christ'. The next challenge is to define what constitutes the 'work of Christ', or the purpose of the Church.

### *6.2.3 Purpose of the Church*

What is the purpose of such an embodied community from the perspective of the members of the three congregations? Their official declarations on this issue have already been identified. They include worship, fellowship, teaching, prayer, observance of the ordinances, and evangelism. The aim of this subsection is to describe and analyse how the individual members perceive it, using their corporate statements as a guide. But it should be noted that the observance of the ordinances did not feature at all in the individual members' response to the question of the purpose of the Church. Much of the discussion that follows has already been anticipated in the previous section.

Out of twenty nine individual descriptions of the purpose of the Church by members of the three congregations, worship was mentioned four times. This can be deduced from the following three statements already cited in the preceding section: "The Church can be defined as a body of believers gathered together *to worship God*"; "the Church is a place where we *come to worship* and relate with God and to also grow spiritually"; and "The Church is a place where different people come together for a purpose: *to worship God*, to praise God" (emphasis added). Here the object of worship is God and the goal of worship is spiritual growth.

Fellowship as a purpose of the Church was mentioned only once by an elderly mother in First Baptist who says, "The main purpose of the Church is for believers to come together in one faith for fellowship." In terms of frequency, among the officially stated purposes of the Church, teaching featured more in number (7) than the rest in the individual contributions of the members. An example is the following contribution by a male university student in Hilltop:

> For me, the purpose of the Church is to heal the frustrated ones, to teach the people how to use the manual, the Bible, on how to correct their emotions, how to reach out to their inner heart, how to relate more to God, how to know about the Father the more and how to correct themselves when they are wrong and how to encourage themselves on how to grow higher.

Also a deaconess, who is a retired civil servant, and a recent male university graduate in Faith Baptist hold the same view when they respectively say, "The purpose of the Church is to teach and train people in righteousness and to reprove wrongdoers"; and

> As one of my brothers has said, evangelism is our responsibility; but it is not enough. Teaching is also important. We need to encourage more teachings in the church. For example, some people come to church but they do not even attend the Sunday School; we should encourage them, especially our children at home, that they should attend Sunday School because that's where we teach the Bible. On the youths, I want to say that the church should teach them morally and physically.

The stated objectives of the teaching role of the Church from the foregoing are equipping with 'the manual', the Bible, correction of emotions and actions, training on how to relate with God and 'how to reach the inner heart', and training on righteousness and mutual encouragement.

Prayer featured four times in the statements of the members on the purpose of the Church. A mother in First Baptist who is a petty trader puts it this way: "The Church as the body of Christ gathers together to pray that our Lord God will forgive us our sins and help us and maintain peace for us." The fact that the subcategory of prayer appeared just four times in the statements of the members is only a tip of the iceberg on this because, in terms of actual practice and programmes of the three churches, prayer featured prominently as described in the previous chapter in the form of their theology of prayer.

Since evangelism, or soul winning, is the alleged preoccupation of African evangelical Churches by Gifford, it became necessary to test this out. It has already been shown from their official documents that the three churches place emphasis on carrying out 'Christ's commission'. The same emphasis featured in the individual members' understanding of the purpose of the Church. It was specifically mentioned four times in different ways. Two youths in Hilltop respectively say, "The purpose of the Church is to obtain salvation. We are all in the Church to gain heaven"; and "The purpose of the Church is to lead those who have not yet known Christ to know him so that at the end they'd aspire to make heaven and not lose their souls." A middle-aged male member of First Baptist emphatically declares, "The purpose of the Church is to carry out the Great Commission." Nevertheless, this perspective is best encapsulated in the following contribution by a secondary school student in Faith Baptist:

> My own is Matt. 28:19-20. The basic thing there is that v. 19 says that we should go therefore and make disciples of all nations, baptizing them in the name of the Father and the Son and the Holy Spirit. The basic thing in that passage is evangelism, if we go out to evangelise, I don't think all these things that are happening would not be happening.

Not only in terms of statements but also in terms of their programmes and activities, evangelism with the aim of making new converts, indeed, featured prominently in the three churches during the field work. For example, virtually every Sunday sermon from the pulpits ends with an evangelistic invitation for people who wished to receive Christ to respond and almost on each occasion there were respondents. In addition, the churches and their various arms regularly carry out scheduled mission and outreach programmes to different parts of their respective cities, as well as rural communities. Faith Baptist goes out once every month around the city of Port Harcourt for witnessing and its various organisations engage in rural evangelism, especially to some villages in Ogoni where it sponsors three churches, with a view to making new converts and strengthening the churches. Hilltop, as well, carries out outreach activities to rural communities in the Niger Delta on a regular basis. In the light of this, it becomes vital to examine their theology of conversion. This is taken up below, and it will form a major part of the evaluation of Gifford's account of African evangelical Christianity.

However, there are two other major blocks of descriptions of the purpose of the Church by some members of the three congregations that do not directly feature in their official statements. The first is portrayed by a youth in Hilltop who is a civil servant when he says, "The purpose of the Church is also to help arrest those who have been taken captive by the devil by bringing them back to God and also making them know that there is hope in life and with faith you can be whoever/whatever you believe you can become." Here, the Church is tacitly and figuratively perceived as a sort of spiritual police man assigned with the responsibility of 'arresting' people under the bondage of the devil and freeing them to become whatever they intend to become. This re-echoes the concept of 'total freedom' of the Pastor of First Baptist, which we shall return to in the analysis of the churches' theology of conversion, and aligns with the African worldview. Such a worldview, according to John Mbiti, is very heavily populated with spirit beings, both good and bad, and with the spirits of the ancestors, with the Supreme Being or God at the apogee (Mbiti 1969, 75; cf. Metuh 1985, 38-46; and Turaki 2007). Suffice it to say that this arrest is supposedly carried out through prayer and the spoken word.

The second additional block of understanding of the purpose of the Church has to do with its exemplary role to the world. This is at two levels, personal and corporate. A youth leader in Hilltop, who is a graduate, represents the personal perspective to this when he says, "My own view is that the purpose of the Church is for us to live in such a way that people would see us as the Bible, to live a lifestyle that would portray Christ." Another youth in the church adds the communal dimension: "The Church is a place where different people come … to live together as a community that shows example to the world." A middle-aged woman in Faith Baptist, who is a medical doctor, stretches this view further when she comments as follows: "The purpose of the Church is to be a light in the world; to be an example of what God is really like; and to react in situations as God would have us react; to bring peace upon the earth; to reconcile people who estranged from one another; to show the glory of God."

When put together, the above descriptions reveal two major sets of understandings of the purpose of the Church, or "the work of Christ," by the members: the spiritual, with 'other-worldly' inclinations, and the social, with 'this-worldly' orientation. The first block comprises evangelism, teaching, fellowship and worship. They are inward-looking and aim at conversion of individuals and preparing them for heaven using the word of God as tool. The second block is oriented towards the outside society and has modelling of godly character and Christian virtue as its goal. This perspective is represented by a member in Hilltop who says that the

Church is made up of people who live together as an exemplar of character to the world, and a member of Faith Baptist who sees a six-point purpose for the Church – a light to the world, godlikeness, peace-making, healing broken interpersonal relationships, and the glorification of God, which is directly related to the Church's ministry in the socio-political realm. Thus, there is some unanimity of opinion in the three congregations that the role of the Church covers the spiritual and the socio-political aspects of life. These contributions were made during the focus group discussions and those present responded with spontaneous clapping of hands, signalling their endorsement of the views. What about their views on the specific issue of Christian social involvement? This is addressed next.

## 6.3 Christian Social Responsibility

The theme of Christian social involvement was of particular interest to this author as a result of one of the key aims of the study, namely, to ascertain whether or not it is true that African evangelical Christianity lacks social responsibility and is anti-development. This issue was specifically raised in the focus group discussions in each of the three churches and the answers were forthright.

First, a university professor, who is a deacon in Faith Baptist, emphatically declares:

> It must be a miserable kind of Christianity that says, 'well, I'm heading for heaven; I'm not part of this world.' God wants you to enjoy this earth even before you get into the kingdom in heaven. So the Church, as far as I'm concerned, has not behaved responsibly in that regard. The Church shouldn't directly get into politics but at the same time it cannot afford to be apolitical, that means, these things that are happening do not concern us. Precisely, therefore, it is the duty of the Church to speak out quite often on these killings, on social injustice, on the inequities in the country. But when the Church keeps quiet, the government will think that it's doing well… The point I'm making precisely is that the Church ought to be fully involved in exposing social evils, criticising them and even telling the government the right things to do.

To this member, there are two kinds of Christianity: one that is so heavenly-minded that it lacks earthly relevance – "a miserable kind of Christianity," and the other that engages with the society by exposing and speaking against social injustice. From his perspective, the Church in Nigeria has not done satisfactorily well in this regard, a claim that was disputed by two other members. One of them, who is a security man in the church, first sang a short chorus in response:

> Tell me what more, what more Jesus can do
> Tell me what more, what more Jesus can do
> Jesus laid foundation and opened the gates of mercy
> What more, what more can Jesus do, tell me what more

And then added, "As far as I'm concerned, the Church has the burden of praying for the society everyday, for a change for the better; but I've discovered that the more the Church prays the more crime rate increases. You understand what I'm saying now. So the Church has done her bit." A voice vote of members of the church present at the focus group discussion (109) on whether the Church has done enough in the region in rebuking evil and injustice was approximately forty percent affirmative and sixty percent negative.

Concerning the situation in the region, some other members of First Baptist and Faith Baptist argue that the Church is to pray as well as take other steps. Two members of First Baptist, one a middle-aged mother and the other a male school teacher, respectively state in the following words:

> As a local church we have a responsibility of praying and fasting and also to preach peace to the most affected persons, especially our youths, to maintain peace and that there are other ways of finding solution to their problem, may be through dialogue. Then CAN as a body should be responsible for most of the work. That is, they should personally visit all the Niger Delta areas preaching the same gospel of dialogue and at the same time taking back their message to Abuja, the seat of power, to tell them the problem the Niger Delta is facing, without fear or favour. In so doing they will come out with a reasonable solution to the problem.

And

> My opinion is that Christians are supposed to do what you're doing, the type of thing you're doing, to go round the Niger Delta and find out the facts that would help bring peace and then dialogue with the Government to maintain peace. Before they do that, they should be conducting prayer and fasting at local and national levels.

The following contribution was made by a male undergraduate in Faith Baptist:

> The purpose of the Church is related to the role it can play in this crisis. The Church and some Christian groups have played some role in this crisis. For instance, the Kalabari Prayer Network has been going to some Kalabari communities that are affected by the conflict to organise prayers and fellowship with and for them. They also use such avenues to witness to them and teach them the way of peace. Also, in the area of prayer, the

> Church has not failed in praying for the communities and for the nation. Another thing is that the Church has been able to preach salvation to members, as well peace and forgiveness… But the area the Church has not done much is that it has not really taken a stand on some issues, for instance, to tell the government that it has not lived up to expectation.

The main points of the two contributors above on the issue of ecclesial role in the midst of conflict are prayer and fasting for peace, direct involvement by visiting the affected areas to gather "the facts that would help bring peace," preaching "the gospel of dialogue", and lobbying of those in authority. Another member of Faith Baptist who heads the church's Nursery/Primary School pushes these thoughts further by adding that Christian churches, at local and/or denominational level, should take practical steps, such as the organisation of rallies and the issuing of memoranda, communiqués and press releases. He comments as follows:

> Yes, the Church has social relevance. With particular reference to the Niger Delta crisis, the Church stands for peace, which is lacking in the region presently. The Church should preach and insist on equity, fairness and justice. In addition to praying, it should also take some practical steps in helping to resolve the problem such as organise rallies, issue memos, communiqués and press releases. It can also play the role of a mediator whereby it will bring together all the parties involved. In addition, the Church should see this thing more as its responsibility than PENGASSAN and NUPENG. So we should wade practically into the matter. Even a church like ours, or the Rivers Baptist Conference, can go to the Rivers State Government, the South-South Governors, or even the Federal Government and make a presentation, raise a memorandum, sorry, a communiqué, on this issue and say this is how far we have gone into this problem. I think that we have a very serious responsibility in the midst of this crisis.

A mother in First Baptist adds as follows:

> The purpose of the Church is not only to evangelise the lost souls but also to engage in philanthropy and in giving assistance to the needy, the motherless homes, old people's homes, in addition to having educational endowment to assist the less-privileged children who cannot afford education except they are assisted. Before now I remember that most of our people who were educated were sponsored by churches. Some of them were even sent abroad to study and, when they came back, they assisted in uplifting their churches or communities. But these days such endowments are no longer available.

She was supported by two other members, one a female school teacher and the other an adult male who, respectively, argue, "My view is that the Church should help the needy and the orphans so as to bring them up to life because some of them are homeless; and some are parentless. The Church should do something for them in order to help bring them up"; and "What the churches should do is to preach against injustice in the country and tell the Government to give the people their rights."

A university student in Faith Baptist added another dimension to the social responsibility of the Church in the region, namely, creation of skills acquisition centres:

> The churches should see how far they can create skills acquisition programmes for youths to empower them for I know that the reason some of the youths are doing these things today is because they are not empowered. So the Church has a very big role to play. I know when the Church came into Nigeria in those days with Western education, I know that they did very well; if they can imbibe the same attitude today by creating room for people to learn skills, I know it will go a long way in bringing the minds of the youths towards doing things that are better in the society, instead of causing problems.

The contributions above focus on Christian involvement in the area of social activities aimed at helping the needy and less-privileged in the society.

The theme of Christian social involvement also featured in Hilltop. For example, a young lady posits in the following words:

> I want to say that Hilltop church is not a church that believes in teaching only the word. We also believe in a lot of social activities. Like the Men's Fellowship and the Women's Fellowship organise different activities outside the church in hotels and halls, whereby non-members of the church are invited so that we can renew our relationship, educate ourselves, play games between the men and the women, and then have dinners together. I also remember that Heart-to-Heart Foundation organised Operation Feed the Children about two years ago. In the last two seminars, Dr. Oarhe brought people to educate the youths on how to start their own businesses and they invited a lot of Micro-Credit Finance companies to come and finance the people. So Hilltop has influenced the society in various ways.

Another member of Hilltop, a youth who hails from northern Nigeria, comments as follows:

> I want to say that the church has been very helpful in the situation in the Niger Delta. I want to use somebody as a case study. I don't know if he is here right now; he's among those who used to strongly believe in taking

> up arms and challenging what is happening in our society today – they want to challenge the Government so as to benefit from what they believe is naturally theirs. But the church has been able to refine his thinking about these things. The person in question was seriously involved in the violence but the church has been able to change his orientation as he no longer believes in carrying arms and doing all sorts of nonsense.

The argument of the above two members of Hilltop is that their church engages in benevolence activities to the less-privileged in the society by way of free distribution of clothes, food, and drugs. They also assert that the church organises a number of activities in hotels and halls whereby it gathers together both believers and unbelievers to educate and equip them with tools for better relationships. Such activities include sport and seminars. Specifically, according to them, in 2004 the church organised a special programme for children through its H2H Foundation. The programme was tagged, "Operation Feed the Children," and through it food items were freely distributed to the children of the less-privileged in the society. In addition, they claim that Hilltop has collaborated with some micro-credit finance companies to teach the youths on how to start their own businesses.

Also, it is claimed that a major role of the church in the conflict is the refinement of the thinking of some militants such that they have renounced violence. A former member of one of the gangs, who is now a member of the church, confirms this as he testifies as follows:

> I used to be a youth leader in Bayelsa State but before I gave my life to Christ, there was a crisis in which I was involved as a fighter. In fact I would have been killed but I came across a lady who worships in Hilltop who invited me to church. At a point while one leg was inside, the other was outside but when the crisis started, Pastor Ben followed me up. I was in the Government House in Yenagoa. What I'm trying to say is that the church has affected me in the sense that if Hilltop were not to be in Port Harcourt, probably I would've been a dead man.

Similarly, a Pastor in one of the congregations in this study mentioned that his church carried out some ministry to some of the boys directly involved in the violence. According to him, through such ministry, some of them have 'repented' and renounced violence and now attend his church. He also relates how four youths at different times came to him to surrender their guns. He says,

> An IYC member came to me to surrender his AK47 rifle and its accompanying ammunition, saying that he was tired of the whole thing. When the boy handed the rifle over to me, we both demobilised it, went to the jetty in the night from where we paddled a canoe to the high seas

> where it was deep enough and threw both the gun and the live ammunition into the sea. The boy in this case now attends the church. It is the fourth time this has happened.

What is the basis of this renouncement of violence and refinement in thinking towards non-violence? The answer could be fathomed from the following submission by a mother in First Baptist: "One thing we should know as Christians is that our Lord Jesus Christ is not a person that believes in violence. So as Christians we should not be violent. We should preach against violence and injustice and then pray to seek the face of God because there is nothing God cannot do." This re-echoes the issue of Jesus and violence earlier noted in chapter five.

There are other aspects of direct social involvement by Hilltop that were noted in the course of the research. They have to do with the condition of the two roads that directly lead to the church location, the issue of the family in the region, and health care outreach programmes of the church. These would be obvious from the following statements by the church Co-Pastor:

> Practically, I don't know about the other churches, it's like they are not interested in the environment. But this church has been doing up the road you're seeing and it's quite a huge task. The church has really helped in other areas apart from leading people to Christ and showing them the way to heaven. For instance, another kind of crisis that I have witnessed in the Niger Delta is not only the destruction of lives but also the destruction of the family structure. With the oil came a promiscuous life style and that has made the Niger Delta most of the time not to have a normal family setting which Heart-to-Heart also helps to put in place.

In the above contribution, the Co-Pastor highlights the fact that it is their church that was regularly maintaining the roads that lead to their church and that they have been involved in addressing the problems of destruction of the family structure and promiscuous lifestyle associated with the oil industry. On the roads, this researcher observed that in-between the two visits he made to the church in September and October 2006, some repair work had been carried out on the two roads with sand and stones to make them a bit passable and motorable. It was during the focus group discussion of Wednesday October 4, 2006 that he got to know that it was the church that had carried out the repair work. Another member who works in one of the oil companies and heads one of the church's departments confirmed this and adds that their local community has written them a number of letters of commendation on account of it. He adds that the church carries out free rural health outreach programmes to rural communities in the region in collaboration with some NGOs. In his own words:

> Yes, the church has been a tremendous and major player in our environment because when we came here as a church, what you're seeing today around was not here. It was when we came here and started setting up this place that people began to see that life can actually exist in this area. Like my mother in the Lord has said, this road you're seeing, we are a major player in it. We are the ones that maintain it. In fact, on every sanitation day, our church members come out en mass to work on the road. It is such that the community has written us several commendation letters concerning it. We also go out for benevolence: we organise benevolence visits to the community. In addition, we engage in community health as there is an arm of the church that goes into rural communities to carry out free medical care whereby we treat the people free of charge. Before we visit any community for the free medical care, we send out information that the Hilltop Medical Outreach would be coming on specified date, time and venue and we've been doing this consistently. We also share clothes and food to people.

The table below gives a breakdown of the church's free health care programmes to rural communities in the region in partnership with the Niger Delta Development Commission (NDDC)/A.M. Projects in August and September 2006:

**Table 2 Statistics of NDDC/A.M. Projects/The Hilltop Rural Free Medical Outreach for August and September 2006**

| | 6th-12th Aug | 27th Aug-2nd Sept. | 3rd-9th Sept. | 24th-30th Sept. |
|---|---|---|---|---|
| **Location→ Treatments ↓** | Opokuma, Bayelsa State | Bomadi, Delta State | Uzere, Delta State | Bonny, Rivers State |
| General Consultation | 958 | 1700 | 1860 | 1925 |
| General Surgeries | 13 | 53 | 46 | 77 |
| Eye Surgeries | 11 | 34 | 21 | 20 |
| Ophthalmic Medications | 184 | 230 | 280 | 259 |
| Glasses | 224 | 186 | 206 | 264 |
| Dental Procedures | 71 | 109 | 209 | 204 |
| New Converts | 212 | 313 | 280 | 562 |
| **Total** | **1673** | **2625** | **2902** | **3311** |

As indicated in Chapter Three, the NDDC was established by the Federal Government of Nigeria as an interventionist development agency in the region in response to the peculiar economic challenges and environmental degradation in the region occasioned by oil exploration. As part of its programme in the area of rural health care, the NDDC initiated collaboration with some Christian Health NGOs. It was in this vein that Hilltop came in as a major player. The church, through its Evangelism Teams, along with its partners, as the above chart shows, goes to rural areas to administer free medical care to the villagers. Some of the treatments cover the areas of general consultation and medication, general and eye surgeries, ophthalmic medication, dental procedures, HIV/AIDS counselling, and evangelistic counselling. These are given free of charge by volunteer Christian doctors, pharmacists, nurses, dentists, pastors and other personnel. According to them, on average, each rural Health Care Mission lasts for a week in each given locality. Thus, between August and September 2006 four of such Mission trips were made to four rural communities in Bayelsa, Delta and Rivers States. Usually, before the medical team for each Health Care Mission arrives, the Evangelism Team of the church goes ahead to prepare the people by way of prayers, songs, and preaching. They also issue cards to those who would eventually see the medical personnel. The villages that were visited during the period include Opokuma in Bayelsa State, Bomadi and Uzere in Delta State, and Bonny in Rivers State.

The above data shows that the church and its rural health care partners attended to the needs of 10,511 patients in four trips. Also, according to this report, not everyone that received medical care became a 'convert'. It is significant to note that according to them, the church does not have a branch in any of these villages and did not see the primary purpose of the trips as church planting but as an opportunity to minister to the physical needs of the people. In the cases where they recorded new converts, such converts were handed over to the pastors of churches in the communities, especially Pentecostal pastors.

## 6.4 Christian Political Involvement

A related issue of interest to this study and which attracted much discussion in the three congregations is that of direct Christian political involvement. The responses were unanimous: Christians should participate in politics. Some added that even Christian clergy should actively participate in it. Already it has been indicated how a university professor in Faith Baptist argued that the Church should neither be apolitical nor directly involved in politics. An elderly deacon in the same church adds another dimension to the issue when he argues as follows:

> I want to say that the Church, especially pastors, should try as much as possible to have a boundary with politicians; not that the Church should not be involved in politics for, if the Church is not involved in politics, how can we have good leaders in the society. We want good leaders, fine. So the Church should be involved in politics; pastors should also be involved in politics.

This view is further portrayed in the following comments by a mother in First Baptist: "Christians should join politics. Even Reverends and our deacons should join so that we can make a difference and also convert those people that are to perish so that they too will come and worship God." A youth leader in Hilltop further argues,

> I think the way the Church can handle or influence the situation is when we directly get into the political scene; it is one thing to complain from outside; it's like when you are watching a football game, if you are not a footballer, you are just clapping... the people that influence the scoring are the players. But if you are not there playing in the pitch, you cannot influence anything. I think like what Jesus said, we are like yeast; we should put ourselves into the game so that we can influence things positively. Christians should participate in politics.

Another youth in Hilltop who is a lawyer agrees with the above submission and links it to Jesus' social ethics and the example of Daniel in the Old Testament. Thus, he says,

> When Christ was here, there was injustice, and the only way he addressed the issue was to teach the principles of the kingdom and how to approach God. He taught about prayer; he taught about conduct in the sense that prior to his arrest in the Garden when Peter, in an attempt to prevent him from being arrested, struck a servant and cut off his ear. Christ said that he who kills by the sword will die by the sword. So violence is out of the question… Christians ought to step into the arena of politics and employ the principles of the kingdom, just like Daniel in the Old Testament. He was in politics; he could interpret dreams even in the area of politics and he found favour with the king and, as a result, he too had a voice…So I think the problem is with the minds of the Christians, most of us think politics is dirty because of the picture the world has presented before us about politics. We ought to re-define what politics should be.

However it should be noted that the low regard for politics is not peculiar with the Christians. For example, Jeff Haynes acknowledges the general aversion for politics by the average person in the African continent when he says, “For many ordinary Africans, ‘politics’ is something to be kept at arm’s length as far as possible. It is something rather unsavoury, connoting the often dubious goings on between elite groups” (Haynes 2004, 80). Thus the aversion does not arise from outright hatred for politics but is as a result of feeling of frustration arising from years of patrimonial politics played by the elite that have produced nothing substantial in terms of development in the continent.

The Senior Pastor of Hilltop agrees with the submission above and adds that leadership development, following the example of Jesus, is the principle challenge facing the Church in the region. Thus he avers in his own words:

> I believe that one of the major things that Jesus focused on while he was here on earth was to raise leadership. And when he left, what sustained his movement, so to speak, was the leadership he developed. I believe that the principle thing the Church can do to arrest the situation in the Niger Delta is to invest heavily in leadership development and training.

Since some of the pillars of the perceived injustice against the region as claimed by the members are the laws that regulate the oil industry (see the next chapter), a former president of the Youth Fellowship of Faith Baptist suggests that the Christian Association of Nigeria (CAN) should lobby Christian members of the National Assembly to amend such laws. According to him:

> CAN should also be able to call together all members of the National Assembly who are Christians and tell them that in the sight of God, what is going on in the Niger Delta is a sin; that this people are producing so much, yet they are the worst hit. When you come to Nigeria they are the poorest. God would not allow this. You people should be able to champion the amendment of all these laws that are impediments to the empowerment of the Niger Delta region, to the region getting their fair due.

The basic line of support for Christian active involvement in politics is that if the Church is to make any positive impact on the society, then Christians should actively participate in it rather than staying afar and condemning what goes on in the political arena. Doing so, they argue, would require the discarding of their aversion for politics which is generally regarded as a dirty game. Hence they insist that not only should churches teach their members that being a Christian and participating in politics are not mutually exclusive, but that Christians should actually go ahead and participate in it. By so doing they would be able to influence the situation from within. Also, the Church in Nigeria through its umbrella body, CAN, it is argued, should engage in political lobbying of the law makers with a view to amending some laws that are perceived to be at the heart of the crisis. Hence, lobbying by church leaders is seen at both the social (as described earlier under 'Christian Social Responsibility') and political levels. However, they argue that Jesus' approach of confronting injustice by teaching on the principles of the kingdom of God is the model Christians should follow. Moreover, just as Jesus emphasised on leadership development, the Church should also invest heavily in developing and training leadership for the nation.

## 6.5 Theology of Conversion

In the light of the foregoing discussion on how the three congregations perceive the Church and its meaning and purpose, as well as Christian socio-political involvement, it is now pertinent to examine their understanding of conversion. This is mainly because their views on these themes are intrinsically founded on an orientation towards the world in relation to Jesus and the Church. At the core of this orientation is the belief that there is something found in Christ through the Church that the world lacks and that the way to be a part of that 'something' is conversion. Also, it became clear from the research that an understanding of their theology of conversion is sine qua non to a full appreciation of their attitudes towards the society and politics. As the following discussion on their theology of conversion would show, that thing that is found in Christ is multifaceted.

A close examination of the views of the members expressed through songs, prayers, and statements shows that there are at least six inter-related meanings and dimensions of conversion. First, conversion is understood to mean a change from their personal and collective past to a new life of "total freedom" in Christ that confers a new identity on the believer wherein he/she transfers ultimate allegiance to Christ whom he/she accepts as Lord and Saviour. The Pastor of First Baptist puts this into perspective in his explanation of "total freedom" which was highlighted in chapter four. In this pastor's usage 'total freedom' is synonymous with salvation and touches on all areas: cultural, moral, social, economic and spiritual. It is in the context of this radical change in identity that their emphasis on being born again should be understood. This is the *ontological aspect* of conversion in their theology. Marshall-Fratani notes this dimension of conversion in Pentecostal thinking when she remarks as follows:

> Conversion to pentecostalism expresses a model for the construction of identity. The notion of being 'born again' encapsulates a particular attitude towards agency and social change, in which the individual is exhorted to make an absolute break with his personal as well as collective past. Upon 'giving his life to Christ,' he re-enters the world as a 'new creature,' as a sort of 'blank slate' upon which the new identity will gradually be written, following a model of spiritual growth from Christian 'babyhood,' into full 'adulthood'(Marshall-Fratani 1998, 285).

Second, conversion for these believers refers to a transformation to the pattern of life seen in Christ Jesus and his teaching/ministry that should manifest in personal and corporate conduct. This is related to the purposes of the church identified by members of the churches. In this regard it is the *moral/behavioural aspect* of conversion. In other words, they believe that the surest testimony to the efficacy of the power of the Gospel and the clearest evidence of conversion is life that approximates to the standard set by Christ.

The *political aspect of* conversion is the third dimension. In their understanding, conversion is an ushering in to the course of social justice and peace brought about through non-violence (for most) or, if need be, through violence (for a few). The majority opinion in the three churches is that Jesus in his life, ministry and teaching stands for justice for the oppressed and is a people-loving, person-centred Lord and Master who is non-violent in his approach in 'fighting' injustice. This could be adjudged from the comment made by a mother in First Baptist quoted in the previous chapter in which she says that since Jesus preaches love and peace, he would not endorse the taking of another person's life; rather "he would encourage us to be our brother's keeper." The missing link here is the absence of a theologically coherent means of relating the micro implication of this aspect of conversion to the macro level. For, as Gifford rightly argues:

> African Christianity today has to take into account both the personal and the social elements. There has to be a double analysis. Systems require change over and above the change in the hearts of those who make them up. This awareness affects the notion of conversion. Conversion can no longer be seen in reference to one's personal sin only. There must also be a recognition of and a turning away from the social dimension of sin, present in the collectivities to which one belongs (Gifford 1990, 15).

Justin Ukpong similarly argues that evangelisation cannot be apolitical because, following the example of Jesus as seen in his ministry, it has "implication for those who govern, the governed, and the way the governing is done, and vice versa" and therefore "has something to do with politics" (Ukpong 1995, 7). Therefore, from the investigation, the problem with the political dimension of their theology of conversion is the *how* and not the *whether*, which is similar to Yoder's political assessment of Pietism when he says, "If ... there was something wrong with the political ethics of pietism, it must have been in the *how* and not the *whether*" (Yoder 2002, 85).

In the fourth place, conversion to them means an ongoing process of inner transformation. Conversion is not just an event but a life-long process of spiritual metamorphosis which begins at the moment of acceptance of Christ and continues throughout the believer's life (cf. Connolly 1993, 159-166). The tool used for the conversion is the Gospel, also called 'Christ's Commission' and 'the Great Commission', and the process by which the transformation is effected is discipleship. Such transformation is made possible through teaching in which individuals are brought to "a higher level of understanding of God." Marshall-Fratani is therefore right when she acknowledges that in Pentecostalism "Conversion is less an event than an ongoing process whose underlying structure is linear and teleological" (Marshall-Fratani 1998, 285). This is the *spiritual aspect* of conversion.

The penultimate aspect of conversion in their conceptualisation to be highlighted is the *existential*. Here conversion is seen as an ushering into the realm of power that helps the believer to overcome the deleterious forces of darkness that held them down in the past and their menace. This is partly the basis of their faith in miracles and signs and wonders and to a degree informs their strong emphasis on prayer. It also derives from their unflinching belief in the continuing efficacy and relevance of the gifts of the Holy Spirit as recorded in the New Testament, which promotes ecclesial egalitarianism and power redistribution, and resonates with the African worldview. Ogbu Kalu alludes to this when he notes that being born again from Pentecostal perspective "offers a release from the forces that dominated one's past as the Holy Spirit flows through from the *pneuma* into the *psuche* and *soma* with power to reestablish the proper relationship and control of God." Continuing, he adds, "Pentecostal conversion was a process of power encounter and resocialization that reconstructed missionary ideology and brought back a concern for felt needs such as healing, exorcism from demonic possession, and similar interests" (Kalu 2008, 252). This should be factored against the backdrop of Gifford's assertion that "It is widely evident that there has been a move away from egalitarian tendencies to a more authoritarian ethos," and that the "'Big Man' syndrome is the curse of Africa, and most Africa's new churches replicate rather than challenge the dysfunctional structures of patrimonialism and unaccountability that characterize Africa's political systems" (Gifford 2004, 175).

Deriving from this understanding of conversion, prayer serves as an existential moment when the powerless tap into power and thereby seek to (re)order the affairs of the powerful through supernatural power. This is seen in two types of prayer observed in Faith Baptist, namely, prayer as prophecy and prayer as release, which were highlighted in the preceding chapter. In the former the members make declarative statements concerning a situation which they believe have the efficacy to bring about the desired result; while in the latter they presume to be the agency of change via their words. What gives them the impetus to do so is their unflinching belief that becoming a Christian and becoming a soldier or warrior for Christ against demons and witches are the same.

Finally, conversion is seen as a salutary alteration of their eternal destiny and assurance of heavenly reward for the believer. This is the *teleological aspect* which defines the ultimate goal of salvation in their understanding. Still deriving from the idea that conversion is a process, they believe that the journey that begins at the moment of acceptance of Christ culminates with arrival at the celestial city, an experience to be consummated with the personal return of Christ. They believe that there are two alternative ultimate destinations for all humans, either heaven or hell, and that each individual's destiny is determined by his/her response to the message of the gospel. Those who reject Christ will perish, or end in hell, while those who receive him will end in heaven where they will also receive some reward. The hope of such heavenly reward serves as a big incentive to their faith and cushions the negative impacts of the diseases, deprivations and disappointments of this life. In the light of this, they hold that they are strangers here on earth and look forward to their heavenly home. This could be seen in the wordings of a song sung by these believers:

> My home is beyond the sky;
> My home is beyond the sky;
> I'm a stranger here;
> My home is beyond the sky.

Their being strangers is defined in diplomatic language as they see themselves as representatives or ambassadors of Christ. As his ambassadors, they are here to represent him.

## 6.6 Summary

The Church as a category elicited extensive discussions in the case studies, especially regarding its meaning, purpose and socio-political role. A number of ecclesial imageries featured, such as body, city of refuge, ambassador, or representative of Christ, and congregation or group of baptised believers, with the body metaphor having the most occurrence. In summary, their ecclesiology could be outlined as follows: first, they emphasise four issues regarding the Church, namely, embodiment, community, space and functionality. Second, they believe that personal commitment to Jesus is the prerequisite to becoming a member of the Church. Third, they hold the view that the Church does not live for itself but is meant to be Christ's ambassador or representative. And fifth, they believe that being a member of the Church carries the obligation of implementing the "work of Christ."

Two major blocks of understandings of the 'work of Christ', or the purpose of the Church, were then identified: the spiritual, with other-worldly orientation, which includes evangelism, teaching, fellowship, and worship; and the social, with this-worldly focus, which includes modelling of godly character and Christian virtue to the world by being "a light in the world" and "a community that shows example to the world."

On the specific issue of Christian social responsibility, they identified the following as the social roles of the Church: prayer and fasting for peace, preaching "the gospel of dialogue," lobbying for justice, organisation of peaceful rallies and demonstrations, and the publication of communiqués, memoranda and press releases. In addition, members of the three congregations identify acts of philanthropy as a major social role of the Church. The researcher also found evidence of road maintenance and extensive free rural medical outreach programmes in one of the congregations. Sequel to the above is the views of the churches on Christian political involvement. There was unanimity of opinion that Christians can and should actively participate in politics so as to be able to change the situation from within. A specific political role of the Church identified is leadership development and training following the example of Christ.

Also, there was evidence of emphasis on evangelism for the purpose of conversion; and conversion is understood as a multilayered lifelong process encompassing all facets of life. Thus, it is not true that they are so otherworldly that they lack this-worldly responsibility as demonstrated by their doctrine and praxis regarding the Church. Neither are they apolitical nor opposed to development as claimed by Gifford.

# CHAPTER SEVEN

# CRUDE OIL AND CONFLICT

## 7.1 Introduction

This chapter examines the prevalent praxis of the three congregations in our study within the context of Nigeria's political economy of oil and conflict. The intention is to ascertain whether or not it is true that they are unconcerned about the plight of the wider society and encourage passive and fatalistic acceptance of misfortunes as the will of God. This will be done by describing and documenting their experiences, understanding, response and witness within the context of the conflict based on the predominant themes determined by word count that emerged from the data under the categories of crude oil and conflict.

The major themes that emerged from the data under these two categories are as follows: crude oil – its ownership and status; conflict – its causes, key parties, experience, impact and response; violence versus non-violence; resource control; Oloibiri; and marginalisation. The following table shows the frequency and distribution of the above themes:

**Table 3 Frequency and Distribution of Themes on Crude Oil and Conflict**

| **Church →<br>Themes<br>↓** | **First Baptist** | **Faith Baptist** | **Hilltop** | **Total** |
|---|---|---|---|---|
| Crude Oil | 32 | 63 | 28 | 123 |
| Conflict | 28 | 30 | 10 | 68 |
| Violence/Non-violence | 42 | 14 | 10 | 66 |
| Resource Control | 8 | 6 | 9 | 23 |
| Marginalisation | 4 | 6 | 6 | 16 |
| Oloibiri | 3 | 4 | 3 | 10 |

## 7.2 Crude Oil and Its Ownership

A major issue that elicited passionate and heated debate among members of the three churches under this subcategory is the ownership of the crude oil in the Niger Delta. Three views were expressed by members of Hilltop. The first view is that it belongs to the Niger Delta people. This is represented by the following emphatic statement by a university student: "It belongs to the Niger Delta people." About one-third of the members (100) contacted through focus group discussions and personal conversations initially supported this view. The second view, which was expressed by a youth, says, "The entity called Nigeria owns the oil." This was at first endorsed by about one-quarter of the members contacted (75). However, the response that attracted a spontaneous debate among the members was the one expressed by a young lady who espoused divine ownership of oil. She declared, "The crude oil belongs to God only." Upon her assertion, some members disputed with her; but overall, the majority supported her. Then, when she reminded them of the declaration in Scripture that "the earth is the Lord's and the fullness thereof," there was a massive shift in position on the part of those who had argued that the region or Nigerian Government owns the crude oil as everyone present agreed with her that the crude oil in the region belongs to God.

The majority of the members of Faith Baptist present at the focus group discussion of August 2, 2006 as at the time this topic was discussed (120 out of 150) were of the opinion that ultimately the oil belongs to God, since he is the one that created it. This was determined by the raising of hands. The following remark by a young lady who is a civil servant encapsulates this view: "As far as I'm concerned, looking at it from what the Bible says, the oil found in the Niger Delta belongs to God. After all, the psalmist says that the earth is the Lord's and the fullness therefore." However they added that God created it primarily for the benefit of the people of the region.

Among the members that held a contrary opinion in the church was a lawyer who argued extensively as follows:

> Yes, God owns the oil; he created it and deposited in the Niger Delta. So, naturally, since we are not yet in heaven, the Niger Delta region owns the oil. But, unfortunately, the Government of the day says that the oil belongs to it, not even to God, and not to the Niger Delta region ... My brothers and sisters, unless we take a holistic approach to the problem of the Niger Delta, we are not going anywhere. The laws are all against the Niger Delta. Section 1 of the Land Use Act says that the land belongs to the Governor and he is the one who decides whom to grant a Certificate of Occupancy. The Petroleum Act says that every petrol, every mineral oil

> belongs to the Federal Government and the Pipeline Mining Act says that when they come to your land, they will pay you either ₦1 or 1k per tuber of yam and you can't claim anything more than that. So, no matter what you do, the law is against the Niger Delta people. That is the problem we're having. And when the oil companies come, ordinarily, what they are supposed to do is what they call Corporate Social Responsibility. If they come to your place and they have made gain, they can say, 'Okay, let's build school for you'. It is not legally binding on them to do it because they pay their dues to the Government which ought to do something for the community and, unfortunately, the Government is not doing this. And when you hold them, you discuss with them and say, 'if you don't do this, you are not coming into our community and everything'. They enter into what they call MoU wherein they state what they will be doing for your community, such as the granting of scholarships, but that MoU is not also binding. It's not a legal document you can go to court and enforce because it does not have the ingredients of a contract, and that is why most of the time they say that the problem is MoU. The companies are violating it and the boys will go back and shut down the oil well or do anything because they know that if they come to court, the court would tell you that the document is not a binding one. So the crude oil, if I'm to answer 'yes' or 'no' in this discussion, is owned by the Federal Government of Nigeria.

The kernel of the above member's submission is that in as much as God is the ultimate owner, yet, at the human level, since "we are not yet in heaven," the Niger Delta region 'naturally' owns the oil. However, this natural ownership is obviated by the fact that through some laws of the Federal Republic of Nigeria, such as the Petroleum Act of 1969 and the Land Use Act of 1978, the Government appropriated the ownership of all mineral resources and land in Nigeria to itself, thereby depriving the people on whose land the resources are located the benefits that should accrue to them. To him all the laws are against the Niger Delta, and if the problem is to be resolved, such laws must be amended. In the case of First Baptist, most members contacted (95 out of 105) endorsed the following assertion by a middle-aged man: "The crude oil belongs to the Niger Delta and nobody else."

In summary, three major perspectives were expressed in the three churches on the ownership of oil. They are natural, theological and legal. The natural view on crude oil ownership says it belongs to the region; theologically it is perceived to belong to God, since 'the earth and its fullness' belong to him; while legally, based on the laws of the land, the oil belongs to the Federal Government. Among these three perspectives, the natural view predominates in First Baptist; the theological view holds numerical sway in both Faith Baptist and Hilltop; while the legal perspective is understood to be the reality which needs to be amended so as to transfer legal ownership from the Federal Government to the region. This issue is related to the sub-theme of resource control, which is discussed below.

## 7.3 Crude Oil: Curse or Blessing?

A major topical issue in the discourse on crude oil among the people of the region is whether it is an asset or a liability. This is usually discussed using two antithetical theological concepts, curse and blessing. This is also related to the debate on resource curse among economists as highlighted in chapter three. Hence, the aim of this subsection is to present the members' perceptions of crude oil, whether it is a blessing or a curse.

The first contributor to this issue in First Baptist was an elderly woman who, speaking through an interpreter, says, "God brought this crude oil to us as a blessing, but now it has turned to a curse"; while a mother adds, "It's a blessing to us." Other contributors respectively said: "The oil is a blessing" (a middle-aged mother); "Crude oil that God gave to the Niger Delta is a blessing to the region, but due to our President or those people in authority it has become a curse to the Niger Delta. If not, God does not make it to be a curse to the Niger Delta" (an adult male); "This crude oil is a blessing to us but we cannot make use of it" (a mother); "The crude oil is a blessing that God has given to us and, now, because of it we have been suffering and it has turned to be a punishment on us" (a middle-aged woman); and "The crude oil is a blessing to the Niger Delta, but the people at the hierarchy have turned it into a problem for the Niger Deltans" (a middle-aged man). However, the contribution that best represents the opinion of the majority of members of the church is the following statement by a self-employed middle-aged man:

> I want us to just refer to two basic things. The first, oil was found in Oloibiri in present-day Bayelsa, but today, how is Oloibiri? Nothing grows there; there is no tangible activity that is vibrant in the place due to oil exploration. Second, we in the Niger Delta area live a very expensive life. What I mean is that the cost of living in the Niger Delta as a result of oil exploration activities is very high. If you look at the multinational oil companies that exist in the Niger Delta non-Niger Delta indigenes occupy most of the high positions, thereby making the Niger Delta people to beg hand-in-mouth without even being employed in these companies and the way the non-indigenes live their lives is far much higher than the indigenes. So the oil found in the Niger Delta today has turned to be a curse for the people of the region because the standard of living is too high for them to attain.

The first response to this issue in Faith Baptist to be mentioned is from a youth who works in one of the oil companies. He argues as follows:

> I want to look at the issue of whether it's a curse or a blessing. It is a blessing but there are two basic problems we have and that is policy and resource management. These are the two basic issues. Policies are from both sides, on the part of the Government and on the part of the oil companies. The oil companies do not keep to the rules because they have the Government standing as a shadow, a backup for them. In fact … there are very glaring cases like the Umuechem case, the Odi crisis, the one close to Elele that has to do with Shell, the one at Andoni that had to do with Gas project and the rest of them; there are so many of them. On policy, Government has not been able to define what should be and how it should be and, where the policies are, they are very weak. It's just like a toothless bulldog. The Government is trying to monitor itself; a man who tries to check himself cannot do that. That is one problem. The other part is the companies. I wouldn't blame them much; they take advantage of our ignorance and because they have the backing of the Government. The other area is the resource management. In fact, this is the most painful aspect. In environmental studies we talk about sustainability of resources or sustainable development. That means, eat today and remain for the future, that is, those who are coming behind. But is this practice being kept? It is not being kept. So what we are trying to do, what we are saying is this: if we are able to manage these resources properly, with good policies and implementation, not just the policy, then it would be a blessing. But when these policies are not put properly, it would become a curse where we begin to see ourselves fighting and killing each other. I think that is the problem.

Another member, who is a retired oil company worker, looking at the issue from the perspective of revenue allocated to the states in the region from crude oil sales, argues as follows:

> If we look at the statistics, that would help us to answer the question: the Federal Allocation that comes every month, Edo State in the last allocation had about ₦3 billion; Delta State has ₦8 billion every month; Rivers State ₦11 billion; Akwa Ibom ₦6 billion; and Cross River less than a billion. But if we look at other States, like Anambra, they have about a billion, Imo about ₦2billion and so on and so forth. So if you look at the quantum of allocation, you can easily say that it's a blessing. The Niger Delta has something on which they can fight for. It would've been terrible if they haven't anything to hold on.

One of the Associate Pastors of the church sees the issue from a theological angle and comments thus:

> I see it also as a blessing, a blessing in the sense that God doesn't make mistakes in his gifts to mankind. So, for us to discover oil in our area, the Niger Delta, it's indeed a blessing. The way a blessing (is) utilised will go a long way in telling people whether we are using it as a blessing or we are turning around to make it a curse or a problem to ourselves. It is the usage now, since the Niger Delta in a way is not being blessed or not benefiting, so to say, in the way it is expected, and many other things are creeping in here and there and people are dying. Therefore people are tempted to see it as a curse. But the truth is that it is a very big blessing to both the Niger Delta and the nation as a whole.

Another Associate Pastor argues that the blessing of crude oil is not limited to the region or the country alone, but to the rest of the world. Hence he remarks as follows:

> I want to tow the line of other speakers. I want to say that it's not only a blessing to Nigeria but the entire world. You'll agree with me that when we have crisis in the Niger Delta oil price will shoot up either by two dollars or three dollars. It shows that it's a blessing not only to us in the Niger Delta but the entire world. But the challenge that we have that is making it to become a problem can be likened to what happened in Genesis. The Garden of Eden was very fine; it was a blessing to Adam and Eve but because they missed it somewhere, it became a problem and they were driven away by the Lord but thank God for what he did to bring them back. So when we look at our own situation, we have a responsibility as a people to put policies in place that will help us to tap the blessing. But one of the biggest problems we have is that even those people that are sitting there, our leaders, don't even know what a barrel of crude oil costs; they don't have the necessary information; the information is there but whether they are not aware or are not using them, I don't know. I want to say that some of the oil companies have done their own part in the past, but what happens is that our chiefs marry more wives; they do some things with the money that they get and that's why you find that, sometimes, young people take cutlasses to either drive the king or chiefs because when they hear that they've collected money and are not using it for the benefit of the village; they go on rampage. So I want to say that it's a blessing, but all we need to do is that fathers in the land, traditional rulers, Government, all of them should put their heads together and come out with concrete policies, not watery ones... They should come up with concrete policies so that this blessing can really get down to the grass root.

On the first issue, through a poll of members conducted, about seventy five per cent of those in attendance at the discussion session were in agreement with the view that it is a blessing to the Niger Delta but which has been mismanaged to the detriment of the region and its people. Not only that, it is also a blessing to the country and the world.

As a result of the problems associated with it, members of Hilltop were divided in their opinion on whether crude oil is a curse or a blessing. For example, a youth who hails from an oil producing community argued from the perspective of the deleterious environmental impact of oil exploration with specific reference to acid rain and says:

> The crude oil in my village has become a curse, even though it's supposed to be a blessing. If you (put a) roof over your house now in my village in Akwa Ibom State, just about three months later, the roof would be corroded because of acid rain that is coming from the sea, and some people are telling us that the oil in the water belongs to people in Abuja. So it has virtually become a curse.

Similarly, a member, who is a legal practitioner, sees oil as a natural endowment whose exploitation has brought negative consequences to the people. Thus, he says:

> Crude oil is a natural endowment. Its exploitation has had negative impact on aquatic life. It also leads to the (emission) of poisonous gases into the air, which the people inhale to their own detriment. From that perspective then, crude oil could be said to be a curse. Oil spill and delays in cleaning it up also add to the problem.

In addition, a youth in the church, while using the specific examples of two oil-endowed communities, namely, Oloibiri in Bayelsa State and K-Dere in Gokana, Rivers State, emphatically states that in his own view, oil is a curse to the people. The following are his words:

> I have to submit that crude oil in the Niger Delta is a curse to us. I want to state that the first oil well was struck in Oloibiri in Bayelsa State in the year 1956, but anybody who cares to know should go to Oloibiri, as I talk now, and see whether the oil that was first struck there is a blessing to the people or a curse... I want to state again that it is a curse, if you go to Bomu in Gokana, or Dere, where oil was struck in 1958, the people are still battling with Shell.

However, there are those in the church who disagree with the assertion that crude oil is a curse to the region. They base their claim on three major grounds: economic, geographical and theological. Economically, a member who is a private businessman adduces the enormous wealth crude oil has brought to the nation to support his position. Thus, he says, "The crude oil in the Niger Delta is a blessing to us and to Nigeria in general because it has brought a lot of wealth to the country. It has become the main stay of our resources; so it is a blessing, though some people, some part of the country has not benefited from it much, particularly the entire Niger Delta area."

Extrapolating further on the above, another youth maintains that by its geographical location, the Niger Delta is in a position to do service to the nation with its oil resource. Therefore he says,

> Oil in the Niger Delta is a blessing. Geographically the position we find ourselves today is a blessing and I want to assure everybody here that God has geographically lifted the Niger Delta people to the position they are today... We are privileged to serve this nation. The crude oil is the source from which every other component part of what we know as Nigeria is benefiting from. I still believe that it is a blessing but, on the other hand, the reason some people see it as a curse is mismanagement.

A young man who serves as an usher in the church adds a theological dimension to support his claim that oil is a blessing. He says, "I want to tell us that crude oil is a blessing because, no matter what happens today, it is a special divine design; so come rain or shine, whether the youths today are ready or not, it is a blessing to the Niger Delta." The rest of the members of Hilltop were divided in their opinion as some supported the crude-oil-is-a-curse position while others agreed with the view that it is blessing.

In all, the topic elicited twenty specific responses from members of the three churches in the respective focus group discussions held with them: eight in First Baptist; six in Faith Baptist; and six in Hilltop. Among the eight contributors in First Baptist, seven held the opinion that crude oil is a blessing, three of which added that as a result of a number of factors, it has become a curse; all the six speakers in Faith Baptist held the same view, with some qualifications; while the opinion was evenly split among the contributors in Hilltop. An aggregation of these reveals that whereas sixteen persons spoke for oil-is-a-blessing, four saw it as an outright curse. Furthermore, about seventy percent of members of First Baptist contacted on this (about 74 persons), seventy five percent of Faith Baptist (123), and sixty percent of Hilltop (242) view crude oil as a blessing from God with a number of caveats. Some of the caveats include suffering, environmental degradation, Oloibiri experience, to which this chapter will return to below, the experience of local people in contrast to that of 'non-Niger Delta indigenes', the negative economic impact of oil exploration as a result of high cost of living, mismanagement of resources, and the actions of 'the people at the hierarchy'. Nonetheless, a closer examination of their positions would show that the two sides are saying basically the same thing: crude oil, which is potentially intended for the benefit of all, has been exploited by the few, thereby leaving the majority impoverished and the environment degraded. To that extent it has become negative in its effect. Hence, those who say it is a curse acknowledge that 'it is supposed' to be a blessing, while those who see it as a blessing concede that it has brought some negative impacts on the people of the region.

## 7.4 Conflict: Causes and Key Parties

The investigation in the three congregations further shows that there is a high level of awareness of the crisis in the region among their members, as would be seen from the following discussion. This is with particular reference to what they perceive to be its causes and main protagonists. Such causes could be grouped into five categories: historical, political and economic, environmental, oil company policies and practices, socio-cultural, and theological.

The historical root of the present crisis in the region was highlighted mostly in First Baptist and Hilltop. This is with particular reference to the country's colonial past. Thus a youth in First Baptist submits, "The problem the Niger Delta is having today is caused by Lord Lugard. If not his amalgamation, this problem should not have happened." (Frederick Lugard was the High Commissioner appointed by the British Colonial Government for the Northern Protectorate between 1900 and 1906. He became the Governor of both the Northern and Southern Protectorates in 1912. In 1914 the Northern and Southern Protectorates were amalgamated to form the Colony and Protectorate of Nigeria with Lugard as the first Governor General.) A middle-aged man in the church claims that it was "stated" before independence that ethnic minorities will be given their own vountry; but this was not done. Rather suppression of the people of the region was intentionally introduced with the fusion of the different peoples that constitute the nation. He spoke passionately as follows:

> From the beginning of politics in Nigeria, in fact before the Europeans gave us our independence, it was stated clearly that the minority should be given their own separate country from Nigeria. Today they have brought all of us together in order to suppress us. Please they should create a country of our own. The Niger Deltans want a country of their own so that they can control their resources.

However, it should be noted that as at the time of the amalgamation in 1914 oil had not been discovered in Nigeria.

Speaking in the same vein, a young man in Hilltop who is a university graduate blames the problem on the "unholy alliance" brought about by the British colonial government when it amalgamated the Northern and

Southern Protectorates and the Colony of Lagos in 1914. Thus, he observes as follows:

> One of the major reasons for the problem is unholy alliance. Before the 1953 Constitution there was the Willink's Commission and you'd agree with me that before this time the structural balance of Nigeria was based on kingdoms. The British people gave us 'unholy alliance'. Hitherto the ethnic groups were controlling their own resources and coming together as individual entities. They didn't have problems with each other, but when the British came, because of their economic interests, they merged all the nationalities without taking into consideration their differences in culture, language, etc.; it became a major problem. This issue was not addressed before Nigeria was amalgamated in 1914. Now the issue has snow-balled to what is happening today. Before this time ... the revenue allocation was based on equity. The regions that produced the groundnut, the cocoa, the rubber were adequately taken care of... Even in the midst of that, the people of the Niger Delta were being colonised in the region, because we were not in the majority and the majority were oppressing us and we didn't have much say. The revenue derivation as it is now practised within the Nigerian context is lopsided because we do not have what we ought to receive, yet we are the people suffering all the consequences of oil exploration and exploitation in the Niger Delta. I want to also submit that the Federation we are running now is lopsided and needs to be addressed.

The major reasons adduced by the member above for the ongoing conflict in the region are colonial policies, especially the amalgamation of Nigeria in 1914; revenue allocation with particular reference to the derivation principle, which he claims is unfavourable to the region; internal colonisation of ethnic minorities by the majorities; the dire consequences of oil exploration activities; and the lopsided federalism being practised in the country. It is noteworthy that there was spontaneous round of applause from the members of Hilltop following this contribution.

The recurring political and economic themes in each of the churches are neglect, marginalisation and discrimination against the ethnic minorities of the region, and the warped federal system being practised in the country. A middle-aged man in Faith Baptist, who runs his own private company, sees it from the point of view of "monkey is working, baboon is eating." He comments as follows:

> You see that monkey is working and baboon is eating. The wealth from the Niger Delta is being totally utilised in the northern part. I've been in the north; I've been in the east; and I've been in the West. In the north hardly do you see black-out; it's only when I got to the eastern part, the Niger Delta precisely, that I saw erratic power supply. So when you look at what is happening, over there in the north you see the drainage, it's as if

> it is like a bridge, a whole river flow of water, but in the Niger Delta there is no drainage; rather you see flood, and people are dying day in day out. The problem of the Niger Delta cannot be counted; it's just numerous and it's only here in the Niger Delta that you see that life expectancy is below may be 60 years.

A former oil company worker, who now runs his own security outfit, describes the Niger Delta as "the goose that lays the golden egg," but which is denied the opportunity of partaking in the benefits of the egg, and likens the experience of the people of the region to that of blacks in Apartheid South Africa. He contributes as follows:

> The issue of the Niger Delta can be likened to the goose that lays the golden egg but is sidelined in the sharing process. Oil was discovered in Nigeria in the Niger Delta in 1956. Over the years, particularly during the military (era), the voice of the Niger Delta people was muffled; they couldn't talk. Again, it can be likened to the issue of Apartheid in South Africa. The South Africans, the blacks, were cheated for years, over 400 years, they couldn't do anything. It's the same thing with the Niger Delta. The Niger Delta people couldn't do anything; there was no crisis but the democracy that has come has liberated them and their voice can now be heard. The fact that you are cheated, the fact that your environment is degraded, does not bring crisis but the fact that you've realised that over the years you have been punished, you produce 90% of what Nigeria boasts of as its wealth, and yet Niger Delta constitutes in the poverty index the poorest segment of this society. It is a pity and unless this issue is addressed there will be no end to the crisis.

To this member, the restoration of democracy in the country came with some irony. According to him, during the years of military rule when there was hardly any space for expression of dissent, the Niger Delta people could barely openly express their displeasure with the injustice they believed were being meted out to them. But the restoration of democratic civilian rule in 1999 suddenly opened the space for agitation. Thus, "the democracy that has come has liberated them and their voice can now be heard."

An adult male, who is a laboratory scientist, claims that the crisis is the outcome of years of neglect by successive Nigerian governments. He puts it pithily as follows:

> Part of the problem of the Niger Delta is total neglect by every successive Government that has been in power. They've been neglected for a long time; and what they do for them is to keep making promises to them of water, light, infrastructure. And how does the Government of the day describe it? They would say that everything is in the pipeline – water in

> the pipeline, light in the pipeline, infrastructure in the pipeline. And because the Niger Delta people are looking for that thing in the pipeline, you see vandalism. So we have vandalism in the Niger Delta today because they are tired of getting those things in the pipeline.

This theme of neglect also featured in Hilltop where a youth uses the cause and effect metaphor to explain it as follows:

> I actually want to see the issue in terms of cause and effect. I see it in terms of neglect which has given rise to youth restiveness. Now, look at the sequence: there was sheer neglect by the Government and the oil companies... A lot of people from the Niger Delta region are now conscious of the fact that they were being neglected and they are still being neglected. This has given them a reason to take a rebellious stance and to destroy oil pipes and do all kinds of things.

On the prevalent federal structure in the country, a young lawyer in Hilltop argues that it is obsolete and needs to be corrected with a view to giving the ethnic minorities a sense of belonging. He remarks thus:

> Indeed, the federation is lopsided; it needs to be addressed. Right from the Republican constitution of 1963, a lot has happened in this country. And I say and I stand to say it anywhere that this federation we are running is obsolete and needs to be re-addressed so that the issue of minority will be balanced so that people can enjoy themselves in this polity called Nigeria.

A middle-aged mother in Faith Baptist who is a lecturer in one of the tertiary institutions in the State ties the root cause of the crisis to the marginalisation of the people of the region on the basis of their 'minority' status, despite the fact that they produce the bulk of the nation's revenue and foreign exchange. She explains in the following words:

> There is no place in the world where people that are discriminated against are happy. Right in our country we're nicknamed minority, I mean the people of the Niger Delta, and because we're called minority we are treated in a minor way, even though the bulk of the wealth comes from here. We are minority, we cannot be heard because our number is few and when the wealth is shared we are given little or nothing.

She also ties the root of the problem to a major political rally that was held in Abuja on March 23, 1998 during the regime of late General Sani Abacha. The rally was tagged the "Two Million Man March." It is alleged that the rally was sponsored and organised by Abacha and his allies to drum up support for his bid to transmute himself from a military Head of State to a civilian President. Youths from all over the country, including those from the Niger Delta, gathered in large numbers in Abuja. The march turned out to be a political irony: a gathering allegedly called to drum up support for Abacha's agenda became 'an eye-opener' for youths from the region who could not believe what they saw in Abuja. They were bedazzled by the level of development in Abuja and its aesthetic beauty and contrasted it with the squalor and underdevelopment that characterised most of their region, the region that they believe produced the money with which Abuja was developed. Upon their return they could no longer hold their peace. Thus she says,

> Our people were not aware on time until some youths visited Abuja and they couldn't believe that that is still Nigeria. They were used to roads with potholes, dry taps, total blackout most of the week, but when they visited Abuja, they were some how upset, knowing fully well that the money used in developing other parts of this country comes from this very slum area that nobody likes to stay. As a result, unrest started.

This same view is shared by several people and community leaders in the region. For example, King Alfred Diete-Spiff, the *Amayanabo* of Toun Brass, Bayelsa State and the first Military Governor of the old Rivers State from 1967 to 1975, said the following words in an interview he granted the *Vanguard* Newspaper on Sunday November 28, 1999:

> Our youths had been comporting themselves until Abacha had his two-million-man march in Abuja. The Niger Delta youths got to Abuja and saw wonderful structures, and said, Oh!, this is where our money is being used, while the government had continued to tell them that there was no money. The youths then believe that they are being taken for a ride. That is why you are seeing what you are seeing today.

Also Daniel Litvin cites this political jamboree organised by Abacha in his *Empires of Profit: Commerce, Conquest and Corporate Responsibility* (Litvin 2003, 259). Thus, the irony of restoration of democracy in the conflict is complemented by the irony of political rally in the days of Abacha.

A subcategory of the political perspective to the root-cause of the problem highlighted by the members is corruption among political leaders in the region. An undergraduate student in Faith Baptist categorically believes that political leaders from the region are responsible for the underdevelopment of the region. He says, "The problems we're having in the Niger Delta do not just come from the Federal Government, but from within us. This is because I believe that some of our representatives at the

House of Assembly go there to share the national cake and forget our communities."

This is further corroborated by a youth in Hilltop who adds unemployment as one of the root-causes of the crisis. He comments as follows:

> I want to specifically mention two basic problems that are disturbing the Niger Delta. One of them is unemployment. It is very, very saddening that most of our youths were left unemployed for a very long time and most of our parents after training us from primary school to secondary school are unable to afford the cost of university education. Even those who have finished their university education cannot get work… And looking for a means of survival, from the creeks they start carrying guns... The other one is our leaders. Here in the Niger Delta, although we complain about neglect, we produce the bulk of the nation's wealth, but our leaders are not doing enough to show that they really feel our problems. All they know is to embezzle the money that is allotted to them as Governors or Chairmen of Local Government Areas… These are the two basic problems: the first one is unemployment, our youths are not employed; and (the) second one is that Niger Delta leaders are not doing enough for their own people.

The two basic factors highlighted above are unemployment and misappropriation of public funds by Niger Delta leaders. These are compounded by poverty, aggravated idleness and frustration, availability of guns, and a fundamental disconnect between the leaders and the people regarding their problems.

In addition to the political and economic factors, some point to the environmental problems associated with the oil industry. Looking at the matter from the perspective of the environmental impact of oil exploration on the economic well-being of the people of the region, a young man in Faith Baptist, who is a civil service union leader, says,

> People who live in the Niger Delta are basically fishermen and farmers. As a result of the oil production activities, there is pollution of the atmosphere and when the atmosphere is contaminated with sulphur and chlorine-related substance, it falls back on the environment as acid rain. When it gets into the river, it kills the diatoms. The diatoms begin the food chain; they are the very small organisms which the smaller fishes eat. When these things are no longer there, the number of fishes that are found in our upper shores is depleted. As a result, the bigger fishes run deep into the ocean and because of this – you know the Niger Delta people are not wealthy people, they don't do offshore trawling, they fish with smaller canoes – when they go to their small rivers, they don't catch any fish. For the farmers, as this acid rain falls, it depletes the nutrients in the soil; so as they farm they don't get good harvest.

It is noteworthy that a number of members of two of the churches in this study who work in some of the MNOCs argue that their respective companies are partly to blame for the situation, by their policies and practices. One of them who is an engineer working in Shell alleges that the company is contributing to the conflict by its 'divide-and-rule' tactics whereby its actions are at variance with its words. He explains as follows:

> Shell is contributing to the conflict in the Niger Delta by its 'divide-and-rule' tactics whereby what it does is different from what it says it will do. For example, the Company recently decided to invest a huge sum of money in the Warri area as a way of placating the restive youths there, while neglecting Port Harcourt area since, according to it, the youths here are more peaceful. As a result, the people in Port Harcourt prepared for a major showdown with the Company, which made it to shelve the plans for Warri.

The crisis is also attributed to the abandonment of the traditional value system that places emphasis on merit, integrity and hard work in preference for "grab, grab mentality," or the "get-rich-quick syndrome." Thus, another member of Faith Baptist who works in Shell claims that many youths in the region are not interested in education or skills acquisition, but rather prefer hand outs from the companies and/or bunkering. He argues as follows:

> Our youths truly are not interested in education or gainful employment. Rather, what they want is hand outs from the companies and bunkering. For example, Shell has a training school in Port Harcourt where youths from various communities are trained in various skills such as sewing, welding, and carpentry. The training is intended by the Company to help equip them with skills and empower them economically but, I can tell you, at the end of the programme, most of them sell off the equipment given to them as a result of the 'get-rich-quick syndrome' since they want quick money.

Furthermore, a group of young boys in Faith Baptist identify some other social factors they believe are responsible for the violence in the region. Two are particularly significant, namely, non-availability of parents for their children and peer group pressure. Thus, a university student alleges as follows:

> Some parents are always busy; most of the time they don't care for their children; they give them money for school, without sparing any time to share with their children, no time to pray with them, always busy doing one thing or the other. So, when the children go outside and move out with

> their friends, they learn bad behaviour from them, since 'evil communication corrupts good manners'.

The Co-Pastor of Hilltop adds some moral and educational dimensions to the cause of the conflict, such as selfishness, corruption, greed, illiteracy, lack of desire for schooling among youths in the region, ignorance and lack of proper leadership as being at the heart of the problem. This could be seen from the following statement:

> I think basically if you want to look at the causes of what is happening in our nation, number one is selfishness, then corruption vis-à-vis greed; and then illiteracy right here in Rivers State that exposed the youths to lack of leadership, nobody to look unto. Let me really express that most of the youths, because of this resource control or the availability of oil in their area, refuse or hardly consider going to school to be educated. If you look at the western part of this country, you would see grandmas and mothers who are well-educated; that's why their youths are all there in the companies. So here, even right now, most of them don't think about schooling but they go about fighting, going to fight this resource control without knowledge. That's why I put lack of good leadership. So the youths need a leader to direct them on how to solve this issue.

However, in addition to the above viewpoints, there are two theological interpretations expressed by two members of Faith Baptist. The first is by a retired oil company manager in Faith Baptist who attributes the problem to the fact that "God is annoyed." According to him, God is annoyed because of human wickedness expressed by way of diversion and utilisation of the resources from oil from the Niger Delta to develop other places, while neglecting the region, in addition to environmental devastation.

The second theological interpretation of the conflict is proffered by a youth who sees it from the point of view of Jesus' teaching as recorded in Matt. 24. In his own words,

> Jesus Christ taught us in Matt. 24, from verse 7, that because iniquity shall abound, the love of many shall wax cold. This is exactly what Jesus Christ is trying to teach us, that a son shall lift his hands against his father; a father will hate his son with real hatred; a mother will abandon her child and go; some will give birth to children, infants, and will just drop them somewhere and leave; this is not even the end yet; that is what Jesus Christ said. It is the beginning of sorrow. So this is one passage of the bible that portrays the experience we are having right now.

In addition, some members of the three churches argue that the extreme form of the violence emanates from the activities of rival cults and/or gangs fighting for supremacy and control of the illegal but thriving oil bunkering business in the area. Again, it is believed that these groups are sponsored by the Government and highly placed politicians who exploit the situation of unemployment by manipulating the youths for their own personal political goals and ambitions. The following is the contribution of an elderly mother in First Baptist regarding this:

> To the best of my knowledge, the people that are perpetrating this problem are the political class, those people that have money; they use the unemployed to carry out this act. During this time I know that there are various groups. One of the groups is known as the Ateke group whose leader is from Okrika; there is another group known as *Dey Gbam* which is led by a Kalabari born young man from Bukuma; then there is another group known as the IYC, i.e. the Ijaw Youth Council, which is led by our own son and brother here, Alhaji Asari Dokubo. These are the groups but their sponsors are mainly the politicians who want to get power by all means. Secondly it's as a result of fight over control of bunkering activities in our high seas.

A young man in the church who is a teacher shares a similar view. According to him, the Government started the problem when it brought together some youths to form 'Vigilante Groups' in the state. He explains as follows:

> As I see the problem, it was caused by the Government because at the initial time they brought together some people and formed them into Vigilante group in the State. After some time a misunderstanding arose between the leader of the group and the Government. So they changed him; but this led to some problem between the two sides and with the Government. It was from there that they started dragging who is more powerful to rule the state. That's where the problem started before it came to IYC and bunkering. Some people wanted to take over control of the oil in the high sea, that is, the vigilante group appointed by the Government. But the (leader of the) other group said, 'No! I'm a son of the soil and the oil comes from my place; so I'm going to take control'. It was from there that they started to challenge each other, which led to the fighting and Government did nothing to stop it as they were supporting one side against the other and bought guns for them.

Indeed a major dynamic of most state governments in the region following the restoration of democracy in the country in 1999 was the formation or adoption of various para-military organisations made up of youths. For example, in Anambra State the Government adopted and strengthened the Bakassi Boys, the Rivers State Government organised the Niger Delta Volunteer Service, while the Bayelsa State Government formed the

Bayelsa Volunteer Force. The membership of these groups is drawn mainly from young supporters of the ruling party which also used them to rig the elections.

Also, according to some members of First Baptist, the Vigilante Group formed by the Rivers State Government, generally believed to be the Niger Delta Volunteer Service (NDVS) led by Ateke Tom, wanted to take control of the oil in the high seas but was resisted by the other group, generally thought to be Niger Delta People's Volunteer Force (NDPVF)/Ijaw Youth Council led by Asari-Dokubo. Consequently, heavy fighting ensued between them as they began to challenge each other and Government failed to intervene. Sometime later, other groups emerged and splinter gangs appeared. As these new groups appeared they aligned themselves behind the two major forces, Ateke's NDVS and Asari-Dokubo's NDPVF. Some of these groups are *Deywell*, *Deygbam* and *Elegam Face*. The alleged leader of *Deygbam* was a youth from Bukuma known as 'Occasion Boy' who was killed in Tombia in 2003 by forces loyal to Ateke. These are three mutually antagonistic cult groups based in three major areas in Port Harcourt, namely, Old Port Harcourt Township, Abonnema Wharf, and Diobu. *Dey Well* group is loyal to Ateke Tom, while *Dey Gbam* members support Asari-Dokubo. The groups later formed female counterparts in 'Red Brassiere' and 'Black Brassiere', respectively. An informant alleged that the primary role of these female groups is to recruit members, especially girls, to join the militant groups.

According to some members of First Baptist, the Government was in support of Ateke's group, whose members were described as being very brutal and mainly illiterate. It is even alleged that they engage in cannibalism. For example, a leader in one of the churches in this study says that on a fateful day in 2004, while he was walking across a street in the city, he saw some human body parts (hands) inside a pot ostensibly being prepared for cooking. It is alleged that this group invaded Buguma City in February 2004, killing, maiming and destroying houses and property. The common belief is that they were sponsored by "a son of the soil," i.e., a native born indigene, who was a top government functionary. On their part, IYC boys, led by Asari Dokubo, came in May same year to serve as a counter force and liberator of Buguma. They were believed to have been sponsored by another son of the soil and also a very senior government official. In the opinion of the church members, "the IYC was used by God to stop the Ateke gang from completely destroying Buguma." In fact a member puts it this way, "Ateke is seen as a destroyer, while Asari is seen as a liberator." To corroborate their view that the Government was in support of the Ateke group in the heavy fighting of

2003-2004, they point out that whereas Ateke boys came into Buguma and stayed for four months unmolested and unhindered by the military, Asari boys came in and within just seven days the military was sent in by the Government to push them out.

Thus, it is believed by the generality of members of the three congregations that militancy is a creation of the state. A pastor in Hilltop best captures this perspective and as well raises some issues for some government agencies:

> I want to say that the issue of militancy in the Niger Delta is a creation of the state. Secondly, it is due to the imbalance in the system. People who ordinarily shouldn't be talking about issue of derivation and related matters have taken advantage of the situation to begin to act as criminals. That's why you see the issue of kidnapping overriding the sense of reasoning which should have been dialogue, consultation, etc... (the) arms they imported and are using passed through the borders. The Customs and Immigration are there; but who controls these? There are establishments that ensure that this thing continues to happen because they benefit financially from it.

On the identity of the militants, four major versions were canvassed in Hilltop. First, some believe they are sponsored by Government functionaries who plan with them to kidnap foreign oil workers for pecuniary gain. To members with this view, such Government officials later share the ransom with the youths when it is eventually paid. A young man alleges, "Most of the youths taking hostages are members of the various militias; some are even Government functionaries. Recently they caught one, a chief of security of the Government of one of the Niger Delta states."

The second version of the identity of the militants says that they are former political thugs armed and used by politicians to help ensure victory for them at the polls but who failed to disarm them after the elections. This is represented by the following submission by a mother in Hilltop:

> Who are those involved? These are boys that were sponsored at the time of election to ensure victory for some sacred cows in this country. Having ensured that these ones entered into power, the weapons were not withdrawn from the boys... So the blame first of all should go to the Federal Government and the laws therein. Then our leadership right here in the Niger Delta, an oil company gets to a place and the leaders or traditional rulers, instead of telling them to develop these boys by sending them to school, would collect money and use it to develop their own families.

Therefore it is argued that militancy in the region is a creation of the state.

The third major version is that the militants are frustrated, 'regular Niger Delta people' who form militias mainly to make money. The following words of a young lady represent this view:

> My view is that most of the people involved in kidnapping people are Niger Delta people. There was a particular day I went somewhere and I was talking with my friends, and a jeep just passed and my course mate said that those are the guys that kidnapped those White men. They are just regular people; you'd never know them and all of them are sponsored by Government; fine, we agree, but some of them just came together. They are frustrated people looking for ways of making money. They go to joints where they capture these people and ask for money. Some of them are not even affiliated with the Government or with anybody; they just come together desiring to make extra money from oil.

The fourth view is a double-version conspiracy theory on hostage taking held by some members of the church. The first proponent of this theory is a young man who is self-employed and he alleges, "The people taking hostages are Nigerians and they live in Nigeria too. I don't believe they are Niger Deltans. At times somebody may want to frustrate Niger Delta people, let's just spoil their name as they've been so blessed." Thus, according to him, the people kidnapping foreigners are not from the region but Nigerians from other parts of the country who are out to discredit and frustrate the people of the region by spoiling their name. Indeed in August 2007 it was reported by Nigerian detectives of the capture of one Afolabi Adebayo, a twenty eight-year old gang member and cultist in Lagos who operates in Port Harcourt but has built a multi-million naira mansion that serves as his hideout in the Ojodu area of Lagos State. According to reports, he confessed that "his marauding gang operates in Port Harcourt and runs to Lagos to relax after making a hit" (See Kehinde 2007).

But a young lady in the church disagreed with the above position by insisting that the majority of those involved are from the Niger Delta, especially Rivers and Bayelsa States. Yet she adds another perspective to the argument, namely, that those involved in the acts are unbelievers. Thus, she asserts,

> I disagree with what my brother has just said. I believe that majority of those doing all these killings are from the Niger Delta. And I think they come mainly from Rivers and Bayelsa States. These guys are unbelievers; they are not believers. If they were believers, they would not go ahead (to) do most of these rubbish things they do.

The second version of the conspiracy theory claims that there is an ongoing "cold war," which is a strategy of "the enemy." It says that the war is ‘cold’ because it is unnoticed and is a ploy to keep the region in an atmosphere of war while the other parts of the country continue to exploit their resources. The proponents of this theory argue that the Government knows exactly what is wrong and what to do but deliberately refuses to do it as it is benefiting from the problem. A mother in the church explains this perception as follows:

> I believe that there is a cold war going on. It’s cold because it’s not noticed, and that’s the strategy of the enemy... I wish I can get it across to the youths to know that it is the plan to keep the Niger Delta in an atmosphere of war while the other parts of the country continue to glory in what they get from here. It is a plan; so I don’t know why the people of the Niger Delta haven’t seen it. The Government knows exactly what is wrong and they are not going to quickly address the issue because they are comfortable with the war going on. If you know your rights, you’d rise up and stand for them; but when they keep you at war, their own side can still be developed while you’re fighting with each other.

In a nutshell, members of the three congregations identify the following as the root-causes of the conflict in the region: the unsolicited amalgamation of the country by the British in 1914, the skewed federalism practised in the country, oppression and marginalisation of ethnic minorities, manipulation of youths by the political class, corruption and mismanagement of resources by those in authority in the region, general feeling of frustration occasioned by poverty and youth unemployment, and proliferation of arms. Others are selfishness and greed among the youths, underdevelopment, environmental degradation, neglect, and unfair share of oil revenue. Consequently they see certain dimensions of the conflict, such as struggle for ‘resource control’, political hooliganism, inter-militia battles and criminal brigandage for pecuniary gains, and inter-ethnic rivalry. As well, the key parties to the conflict identified include the federal government, which allegedly is dominated by the majority ethnic groups; political leaders of the region; local chiefs; oil companies; and the various militant groups.

## 7.5 Experience and Impact of Conflict

Having analysed the views of members of the three churches on the conflict in the region in the preceding section, the next challenge is to consider and document their experience and how it impacted them. This is what this section seeks to do. Evidence available shows that the three churches have had varied encounters of the conflict and it has impacted them individually and collectively.

The first to be examined are some of the key individual experiences as narrated by the members. This researcher was informed of specific members of Faith Baptist who were killed in the conflict. The first was a member of the church from Bukuma who was brutally murdered by one of the gangs in Abonnema Wharf in 2003 while he was on his way to Old Bakana. Before his death, he was a member of the Men's Fellowship of the Church. Bukuma is a Kalabari village which was beset with violent rivalry and struggle for supremacy by various cults and gangs fighting for control of bunkering routes in the area. It is believed that the gang that killed the man was angry with him because they believed he was among those from the community who had reported them to the police. The second was Mrs. Margaret Asawo, a nurse, who hailed from Buguma and who was felled by stray bullets in Port Harcourt from the hands of militants while she was returning from work. This occurred while this researcher was on the field when he went for his second research. The third member was a youth who was extrajudicially killed by Nigerian soldiers belonging to the Joint Task Force (JTF) in Abonnema Wharf in Port Harcourt in late 2008 and his body was never found. This incident occurred while this researcher was on the field during the period of his one year suspension of studies to be with his family as a result of the conflict. He personally made visits to the JTF Headquarters at Bori Military camp in Port Harcourt to inquire about the whereabouts of this youth who was wrongly arrested while he was on his way to church. It took the involvement of the President of MOSOP, Ledum Mitee, for them to own up that the boy was killed while in detention.

In addition to death of some members, several others lost their loved ones, property and/or means of livelihood. A young lady in Faith Baptist who is a university student says in the following statement how her father was murdered in 1994: "My dad was murdered because of the crisis and I can remember for more than three occasions that they've come to my village to shoot and I was in the village and we had to run to the bush." Then a middle-aged mother who hails from Buguma describes her personal loss on account of the conflict as follows:

> My whole town was evacuated as people ran away. There was nobody in my compound. Before the trouble started, my two cousins were dragged out of the house as they were eating. They took them somewhere and killed them. My family house was burgled and everything was taken away and we were just left like that. In fact, up till date, I cannot go to my house because of fear. The whole compound was ransacked. In fact, most houses in Buguma were damaged, looted or burnt down. An old woman, she can't see and her children left her to run for their lives, was burnt alive inside her house.

Some members said that they lost their belongings in Njemanze Waterfront when IYC boys attacked the area on Sunday August 22, 2004. A member who is a chief from one of the Kalabari villages says that he has been living in fear for some years now as a result of threats from some of the militants. He comments as follows:

> Sometime ago my community nominated me and some others to make a formal complaint with the Police Zonal Headquarters in Calabar about the activities of cultists in my village who refused to stop their actions despite several appeals. They were robbing and extorting money from people. Some of them were even involved in raping young girls. So we went. Since then these boys have been threatening to deal with me. They've killed some of those that went with me to lay the complaint. So, my movements are highly restricted. There are some places here in Port Harcourt where I dare not go because the boys are well represented there.

Similarly, another member of the church who is a High Court Judge says that his house in the village was destroyed by a gang of armed youths in 2006. He too has received threats to his life. The following is his account of the circumstances:

> The present crisis in (my village) started in December 2005. It was precipitated by the assassination of a retired Deputy Commissioner of Police, who was a cousin of mine, by a group of people. He was assassinated in cold blood right on his bed. A man from another part of the village had died and a group of people went to consult an oracle in Ogoni on who was responsible for his death. The oracle allegedly said that the retired police officer was responsible. Upon getting the information, they went and shot the officer in his sleep last December. He eventually died in a hospital here in Port Harcourt. As expected, the police authorities swung into action and made some arrests. Partly due to the profile of the officer that was killed, the Inspector General of Police was personally involved in the case. Till date nine of those arrested are still in detention. As a result, some of the people back in the village alleged that I and others are responsible for the continued detention of the nine and they have been threatening to kill me and the others whom they believe are responsible for their detention. In order to underscore the seriousness of their threats, they went and damaged my house in the village and looted my property, including those of my late senior brother.

In addition to the above, some businessmen/women who have contracts with some of the oil companies claim that their businesses have been crippled or seriously hampered by the crisis.

Being located at the heart of Buguma City, members of First Baptist were inadvertently caught in the crossfire of the conflict and its concomitant effects. This is mainly due to the fact that the City was at the epicentre of the crisis between 2002 and 2004. Several cults and/or gangs operate in the town, in addition to the fact that one of the key players in the violent agitation for 'resource control' and the leader of the NDPVF, Mujahid Asari-Dokubo is from there. Consequently, according to them, members of the church suffered in various ways as a result of the crisis. Some lost their lives; some lost their relations; while others lost their property, or all of the above. Upon enquiry twenty six members present at the focus group session held on August 31, 2006 indicated that they had suffered personal losses. Some of the losses include the following:

- A member whose son had registered for WAEC SSCE exam could not write it, despite the fact that she had already paid for it.
- Another one says that her daughter's shop was burnt down and their residential house was burgled. For example, out of a total of 26 crates of mineral drinks left in the shop, 21 were looted.
- A member complained that her debtors ran away during the crisis without clearing their indebtedness to her, thereby causing her some financial loss.
- A mother says that her child who was sick could not be taken to hospital because the hospital was closed down as the medical staff had fled for their lives. There was also nowhere to buy drugs for her son.
- A female member complained that due to the crisis the waterside where she used to trade has been turned into a motor park, thereby denying her of her means of livelihood. As a result she can no longer train her children.
- Another mother says that she ran away for her life during the conflict but, upon her return, she discovered that all her property had been looted. It was only recently that the church through the pastor was able to help her with some money to restart her business.
- One of the mothers said her son was shot dead during the conflict. Also her house was burnt down and a car her son bought for her was carried away by one of the gangs.
- Another one lost her two brothers.

Also, the President of the church Baptist Student Fellowship informed this researcher that students in the church have been affected by the conflict. According to her, several young girls were raped by both the

Mobile Police unit (MOPOL) posted to the town and the NDVS, or 'Ateke Boys'. Her comment is as follows:

> Our church, especially the students, has been affected by the fighting in Buguma. This is because both Ateke boys and MOPOL are involved in raping young girls. I also know that there are some parents in the church whose children belong to the two main opposing militant camps. For example, there is a mother whose son was shot dead by the son of another member who belongs to an opposing gang. The mothers of both of them serve in the same committee in the church. But thank God they can still work together because they know it is not their fault that their children joined cults. In fact, on the day her son was killed, the mother came to church and she said that she had warned her son several times to desist from cultism but he wouldn't listen.

The Pastor of First Baptist corroborated the above and added that between 2002 and 2004 his church was heavily impacted and that at a point in time, more than one hundred members were living as 'refugees' in the church and he and his wife had to provide food for them. Continuing, he adds,

> Some of the armed youths used to come to the church even at the peak of the fighting. They would keep their guns by the door to the auditorium, position one of them to watch over them, and then enter the church for worship. At the end of the service they would pick their guns and go. Similarly, there was a day rumour went round the town that some Ateke boys were inside the church. So some Asari boys, or Ijaw Youth Council members, came in while people were gathered for worship to search for them. But they found none of them and thereafter left without molesting anybody.

Some examples of individual experiences of the conflict in Hilltop include the following: Two young men reported losing their younger brothers to the conflict. Respectively, they say, "The crisis has affected me directly. I lost my younger brother to the crisis. He said he was fighting for the Niger Delta, for resource control, and he was fighting till the Federal Government troops killed him"; and

> The crisis has so much affected my life in a way… in fact I don't want to talk about this again… Sometime last year my younger brother was in the village, he is not a cult member; he was not a part of them, but I was surprised that when they came to raid Buguma as they were shooting, stray bullet hit him and he died.

Also a female member who is a civil servant commented that her eighty-year-old uncle was captured and killed by an Ijaw militant group during one of their raids in Warri and that his corpse was never found. Her words: "I was directly affected too because, by the time the Ijaws came to Warri, my old uncle that could not run, he was in his 80's, was captured and till date he's gone. Praise the Lord! They killed him and his corpse was taken away. Praise God!" (A very significant feature in the contributions of members of Hilltop in the focus group discussions is that most of those who spoke started and/or ended their statements with 'Praise the Lord!' or 'Praise God!')

The experience of another member was not death but detention. According to him, he was the Personal Assistant to the late Ken Saro-Wiwa. He reports that he suffered five consecutive detentions in the hands of the Federal Government as a result of the 'struggle'. Also his home community, Ka in Khana Local Government Area of Rivers State, was burnt down on August 4, 1993, which led to loss of lives and property. He explains: "I want to start by saying that I have suffered five consecutive detentions as a result of this struggle. I was the Personal Assistant to late Ken Saro Wiwa. Added to that, my community was burnt down on the $4^{th}$ of August 1993, and that brought about huge losses in terms of deaths and property."

On his part, the Senior Pastor of the church reports of a personal encounter he had in the hands of some militants. According to him, himself and his Personal Assistant narrowly escaped being lynched by some irate youths along Benin-Sapele road earlier in 2006. This happened when they were returning from his village in Edo State to Port Harcourt in his car when they ran into the youths who were armed with various weapons such as machetes, clubs, and guns. He narrates their ordeal as follows:

> Sometime ago my PA and I narrowly escaped being lynched by irate youths along Benin-Sapele road. We were travelling in my car when we ran into these youths who were brandishing various weapons: machetes, clubs, and guns. They stopped us and some of them climbed to the roof of my car; others sat on the bonnet; while some stood on the side steps of the (4-wheel-drive) car. As these were going on, the youths were shouting at me that I am Itsekhiri and therefore they must kill us that day. I tried to explain to them that I am not Itsekhiri but from Edo State by origin. They refused to listen until my PA and I spoke our mother tongue which was recognized by one of the youths as indeed belonging to an ethnic group in Edo State. After they were satisfied that we are not Itsekhiri, they allowed us to go. This was after they had inflicted some serious damages to my car as they used clubs, machetes and even axes to hit it.

Added to the above physical effects are the psychological ones. Several members of the three congregations say they were living in fear as a result of the conflict, especially those whose houses are located near some of the hot zones. For example, a young mother in Hilltop narrates in the following words her apprehension when some of the boys fought near her house:

> There were times when I and my kids had to sleep outside the house; we had to take refuge somewhere else because of the crisis. In the heat of the serious fighting of the year before last (i.e., 2004), a bullet came into my bedroom; we were not around when it happened. So we are directly affected and, even now, most times when we hear gun shots we have to leave the house because we stay close to a place where, when some of the youths are running, they have to pass through; it is like a creek, so they have to pass through our backyard to escape. So we were affected... My reaction has been fear.

There are also accounts of group experience and impact of the conflict on the churches. A youth leader in First Baptist describes how the violence has affected the church this way:

> Because of the crisis most of our members ran to different places where the crisis did not affect. I remember that before the fighting we used to have about one thousand people in worship but this reduced to six hundred. Because of that most of our members are not in the church, and this has negatively affected our church income as people are not able to give as much offering as they used to and the number of those who give offering has also drastically reduced. As a result, the church is not able to meet a lot of its financial obligations or carry out its activities.

Thus, the church suffered reduction in membership and worship attendance from about a thousand before the crisis to about six hundred due to migration of people out of town, and therefore was unable to carry out its activities as it used to. This is particularly manifest among young people in the church. The youth leader cited above explains as follows:

> A lot of our youths that are supposed to be in the church, the crisis has taken them away to Port Harcourt. They hardly come back. Also, due to the crisis, we cannot see several of our members any more as some of them died and some of the youths went back into the world. Even the "June *Alali*" is no longer as effective as it used to be as attendance at the programme has drastically reduced. ("June *Alali*" refers to the mid-year harvest thanksgiving programme that is held annually in June. It is a major fund-raising ceremony for rural churches in the Kalabari area.)

Looking at it from another perspective, an elderly mother and member of the church Kalabari Choir says that the crisis led to a drastic reduction in the number of its members: "Before the crisis we had so many members in this Choir and when it comes to the time for Kalabari music, you would see a large number of people and you'd enjoy the song."

It was also reported in Faith Baptist that the church collectively was impacted by the crisis. Apart from the loss of members through death, the conflict affected the programmes of the church. According to the members, the attack of IYC boys on August 22, 2004 led to the lowest attendance in morning worship in the church for more than a decade as members could not come out from their homes. Also, evening programmes that usually close at 7:30 pm were closed earlier at the peak of the crisis in 2003/2004 and 2006. It also raised some theological tensions for some members, vis-à-vis their faith in the God of peace when juxtaposed with the violence. The following comments by a former youth president in the church give some insight into this:

> Faith Baptist Church, Port Harcourt has been affected by the crisis in the Niger Delta. I remember last year or two years ago when the killings in the Diobu/water side areas by IYC members intensified, specifically on August 22, 2004, so many of us were unable to come to church and even when we came to church we were always in a hurry to leave, which ordinarily is not our system in this place. I also know that there was much panic among the members in the areas mostly affected. Some members were also displaced and one or two spaces in the church were occupied by such persons during that crisis. So one way or the other we have been affected seriously and apart from that, our faith was shaken at that point because our trust in our God is that he is the God of peace; but having seen so many persons dying, our relatives, friends, many of us questioned the truth of this God we've called out to: how could my brother or my sister die under that kind of circumstances, not a natural death.

This same thought is expressed by a youth in Hilltop, who says,

> The church has been affected in the sense that when we talk about the question of shootings here and there that happened in Port Harcourt, if you want to come to church and you are well-informed and you are told, 'Please there is killing and shooting out there', there is no way you would come to church. We've experienced some scanty attendance in worship and church programmes as a result of the violence.

Yet a mother in the church sees a positive impact of the conflict and submits, "The church has been affected positively: It has increased our prayer life. We've been very much concerned and have been praying individually and collectively due to the problem on ground." The alleged positive impact by this member is the increase in their prayer life.

## 7.6 Response/Witness in the Context of Conflict

The next logical challenge is to probe the responses of these churches and their members to the conflict. Their major lines of response were prayer and provision of temporary shelter for members whose homes were destroyed or attacked, provision of funds for the resumption of businesses, and mutual support and encouragement.

On the provision of temporary shelter for those displaced by the crisis, a youth in Faith Baptist and a mother in First Baptist, respectively, say: "I think giving them tentative accommodation at such a critical time was a serious act by the church"; and

> Actually when the crisis occurred, I was affected and my things got burnt and I ran to the church, to First Baptist Church, Buguma and the church helped me as I stayed in the church for some time. The Pastor and his wife were feeding us. They helped me a lot in prayers. Some of the members even helped me financially so that I was able to restart my business that was destroyed in the course of the fighting. Also we were praying together in order to comfort and console one another.

Also some members opened their homes to fellow members who became internally displaced as a result of the violence. This was especially the case as was witnessed by this researcher in Faith Baptist in July and August 2007.

The personal response of individual members who have suffered one loss or the other as a result of the conflict has been a painful process. To the young lady in Faith Baptist whose father was killed in the crisis, responding to the experience has been a deep struggle between her Christian ideals and her personal loss. The one who lost two of her cousins says that her response was prayer that the violence would cease and that peace would return so that everybody will be happy again. She based it on the conviction that there is nothing God cannot do. Thus, the two ladies above do not see any reason to personally retaliate for the murder of their loved ones as they believe that vengeance is of the Lord.

Members of First Baptist responded in various ways to the violence. First, most of them fled for their lives, leaving a few behind, especially the elderly ones who could not run. The first place they ran to was the church and the first person they called upon was their pastor. Throughout the period the pastor and his wife remained in Buguma to minister to the members who did not or could not run. According to the Pastor, his ministry at the peak of the violence included care, nurture, encouragement and provision of food and shelter for his members as well as non-members.

In addition to the above, the church carried out some ministry to some of the boys directly involved in the violence as some of them 'repented' and started attending the church. According to the pastor of First Baptist, four militants at different times came to him to surrender their guns. The latest of such hand over occurred in September 2006. The account by the Pastor is indicative of certain facts: first, it confirms the extent of availability of guns, especially the AK47 rifle, and accompanying ammunition in the area; second, it shows that there is some element of violence fatigue among some of the militants; third, it indicates the level of confidence the boys who surrendered their weapons to him have on him as a minister; and, fourth, the fact that they went to the sea at night is to safeguard the identity of the boys in question and to ensure that the guns would not be accessible to any other group. When asked why the boys chose to come to him instead of the law enforcement agencies, the pastor explained that the boys said that they did not trust the police and were also concerned for their personal safety.

But this raises two major challenges for Christian churches, the first is ethical and moral and the second practical: First, should former 'militants' who become 'born again' and join the churches be required to account for their roles in the taking of lives and destruction of property, or should there be deliberate amnesia concerning their involvement? Put differently, does personal salvation in Christ Jesus absolve a new convert from liability for pre-conversion criminal conduct? If indeed the quest for justice is the basis of the armed struggle, what about justice for the innocent victims of the violence: the men and women who lost their lives and/or property, the toddlers, elderly men and women, and foreigners who were kidnapped and held hostage, and the general masses of people whose means of livelihood were destroyed? Should the church encourage and/or participate in protecting the identities of former militants who killed and maimed, or should it facilitate their exposure for the course of justice to have its way? For example, Adam and Adam have given an insightful analysis of the six major ways various democracies have dealt with similar situations: amnesia, trials and justice, lustration, negotiated restitution and compensation, political re-education, and truth commissions (Adam and Adam 2001, 32-47). These questions are made more poignant by the fact that "People who believe themselves to be victims of aggression have an understandable incapacity to believe that they too have committed atrocities" (Adam 2001, 44).

The second major challenge centres on the issue of unemployment and poverty. Finding foot soldiers in an army of unemployed, poverty-stricken youths is not a difficult job. The following account by two local pastors on the experience of a former militant accentuates this problem:

> Peter was a member of one of the major gangs in Rivers State. He had participated in various violent activities. One day, through the efforts of an Assemblies of God pastor, he gave his life to Christ and became a Christian. He renounced his membership of his former cult and joined the church. He became very active in the church to the extent that when the pastor wedded, he was his Bestman. But Peter had no job. After a while, he decided to look for job in another village. But the fact was that his former gang had become so powerful in the area that it determined who gets job with the oil companies around the area and, for you to get a job, you must be a member. So, Peter was torn between continuing joblessness and membership in the church and rejoining his former group and getting a job. Out of frustration and desperation, he chose the latter option. So, he approached the group, who laughed him to scorn for deserting them. After recommitting himself to their membership, they linked him up with an oil company which gave him a job. Unfortunately, one day, while on a journey by boat, the boat he was travelling in was involved in a head-on collision with another boat, and he was the only one that died in the accident (For security reasons and at the request of the two pastors who gave this account, 'Peter' is a pseudonym).

The practical challenge before the Church from this story is what it can do to assist in addressing the problem of unemployment within its rank-and-file and in the region. Could the churches either singularly or collectively initiate micro-credit facilities so as to economically empower the poor and unemployed?

Generally the same lines of response recorded in the two Baptist churches were repeated in Hilltop. For example, a member mentioned prayer and the restraining influence of the church in the region. Already their theology of prayer has been addressed. On the restraining influence of the church, the following submission by a youth in Hilltop who once contemplated joining the militancy is representative of the perception:

> The existence of this church has also helped because I am from the Niger Delta too and part of the places the crisis is happening. By the grace of God, if not for this church, through its teachings and ministry, probably, I would have been involved in it too, by carrying guns and shooting. Thus, the church through its teachings and ministry has played some restraining role.

Another youth in the church who is not from the Niger Delta by origin concurs by saying:

> I want to say that the church has in deed responded immensely through sermons and messages, even though they have not all responded but it's just a matter of time. I have seen a lot that has been done towards that. Even though I'm not from the Niger Delta but, so long I reside in the Niger Delta, I'm a member of the region.

The immediate reaction of members of the church upon being affected or receiving the news of personal loss as a result of the conflict is represented by the following contributions by a youth leader and a nursing mother, respectively:

> When the crisis affected me by the death of my younger brother, immediately I heard the news, the first people I informed about it was the church through my head of department and I asked them to pray for me that even where I was, I was not safe. The prayer was rendered. Miraculously we buried him and I came back safely, even if at the end of day I still lost my younger sister as a result of it. I still went back to the church to still pray for me because I know that is where my hope and my strength lie.

And

> My initial reaction when the fighting occurred near my house – I was the only one at home with my very little baby then as my husband had gone to church and everywhere was quiet – was fear because I heard so much from behind my house and I peeped through the window. Before then there had been a lot of gun shots and so much noise. When I peeped through the window, I saw many boys armed with guns coming from the back. I was afraid. So I moved my daughter and called my house help to come in and we lied down somewhere to hide. I tried to call my husband so as to inform him but his phone was switched off. Then I called my head of department and I spoke to her in very low tones and I informed her that the boys were behind. She encouraged me by telling me that it was well, that nothing was going to happen to us and that in Jesus' name we were under a covering. And then I looked behind – I believe it was God because they could have jumped because they were actually running – but they passed by my fence and went to another place. Those boys harassed people around. They went to my neighbour and took her sons' clothes to change because the police was looking for them. So I believe that it was God that diverted their attention from my house and they went somewhere else. I did mention last time that my elder brother works in the field. During that period there was a shoot out when they were travelling to Escravos; his boat was caught in a shoot-out between Okrika people, i.e., Ateke boys,

> and the other gang. There was a serious shoot-out because the other people thought that it was an enemy boat that was coming. Somehow God helped them and they all went under. There was serious shooting. The boat driver was shot, likewise the person that was standing behind him. Fortunately, nothing happened to him.

From the foregoing, the response and witness of the churches in the context of the conflict are fright, flight, and forbearance at the personal level. At the corporate level, it includes prayer, provision of temporary shelter, food and money for resumption of businesses, mutual emotional and spiritual support and encouragement, and as outlet of disarmament and gang decampment. This re-echoes the discussion on the church as a community highlighted in the previous chapter and confirms Ruth Marshall's observation that "the churches see themselves, ideally, as providing moral and spiritual guidance, hope and personal empowerment, as building spiritual and moral communities" (Marshall 1995, 241).

## 7.7 Violence versus Non-Violence

A further theme of interest in this research which featured prominently in the case studies is violence and non-violence. In the light of their experience, understanding and perceptions of the conflict, and in the light of their theology and ecclesiology as described in preceding chapters, the study sought to find out the views of the members on this subject matter. The issue of Jesus and violence from their perspective has already been discussed in chapter five.

Notwithstanding their attempts to explain the reasons for the violence in the region, all the members of Hilltop contacted (343) on this theme through the focus group discussions and private conversations were unanimous in their rejection of violence as a legitimate means of resolving the problem. All those who spoke on this supported non-violent approaches in resolving the crisis and they were applauded by the rest. Speaking from the angle of development, a young man argues as follows:

> I believe there are other routes in solving this problem: the stakeholders involved should be brought together where they can air their views and find a way to solve the problem. By stakeholders I do not mean those in Government but those people that are really fighting, that are causing this conflict. They should bring them together and resolve it in a diplomatic way. There is no development in violence.

The kernel of the above submission is that all the stakeholders involved in the conflict should be brought together to discuss all the relevant issues. Significantly, the speaker excludes Government officials from those to be reckoned as 'stakeholders'. The background to this might not be unconnected to the fact that "'stakeholders' meeting" was one of the major tactics adopted by the Obasanjo administration, especially between 2003 and 2004, in its attempt to douse the tension in the region. Such meetings produced little or no result in terms of resolving the conflict mainly because they were elitist in organisation and participation and therefore did not involve people at the grassroots level (Osuigwe 2007, 35).

While agreeing with the above contributor, a university graduate in Hilltop highlights what he thinks are the major steps that must be taken in the non-violent approach if it is to work: dialogue, consultation, direct dealing with those that are mostly affected by the crisis, and mediation. His words: "I believe there are alternatives to violence... (1) dialogue, (2) consultation, (3) dealing directly with the people that are mostly affected, and (4) there should be mediation. Let the proper issue of federalism be addressed."

Also, while speaking in favour of dialogue, another youth raises some points on the issue of equity and the moral climate of the oil economy. He also avouches that churches should engage in concerted prayer and fasting programmes for the resolution of the crisis. He argues as follows:

> I'm from the Niger Delta. When we talk about the Niger Delta, I talk from the point of view of interest because I'm concerned... How do you expect a people, who are kept perpetually backward, while they see the main stay of the country flow through their backyard, to keep quiet! It is practically impossible to convince these boys to relax, especially when it has taken a very long time and the issue is not addressed... Then the solution of course is dialogue and the whole Niger Delta will agree to face God, set out some days for prayers and fasting, the entire Niger Delta, not just one tribe, (and) wake up (to) prayerfully address this issue.

In the case of First Baptist, out of ten members who specifically spoke on this issue, eight persons spoke for non-violence, while two were in favour of violence. The response of the rest of the members by applause also indicated where their view lay. Majority applauded those that spoke in favour of non-violent approach as against a few that supported the violent mode of agitation. To such members dialogue is the best way, or "the more civilised way." Thus, a middle-aged mother says:

> The best way, or the more civilised way, of settling this issue is dialogue. I feel a round table discussion is preferred to violence in the Niger Delta. I know that the level of unemployment and abject poverty in the Niger Delta

> perpetrated by the previous leaders is one of the causes of the problem. They should create a lot of employment opportunities for the youths. Also educationally they should be encouraged. Then for those people that dropped out from school, they should create other trades like carpentry, welding, etc. This will help to solve the problem.

Another mother agrees with her, by saying,

> It is not through fighting and killing people that we can get what we want. This is because the fighting and shooting has led to the death of so many of our elderly people and some have incurred sickness. What we believe as Baptists is that we have to pray throughout the Kalabari Kingdom, Rivers State and the entire country that God is the only solution who can maintain the peace for us. If we love ourselves we can reason together so that there'll be peace among us. We can find a place, may be in the Niger Delta, and sit down and discuss it, or we go to Abuja. When we sit and discuss it, if we love ourselves, we can have peace.

It is worthy of note that the American existentialist psychologist, Rollo May in his classic monograph on the psycho-social roots of violence, *Power and Innocence: A Search for the Sources of Violence,* also identifies love as a prerequisite for the eradication of violence. He says, "Compassion is the name of that form of love which is based on our knowing and our understanding each other. Compassion is the awareness that we are all in the same boat and that we all shall either sink or swim together. *Compassion arises from the recognition of community*" (May 1972, 251; emphasis in the original).

Yet, two adult males raise the issues of equity and resource control. Respectively they say, "For us to get peace, equity in this our land, the region must be given her right, right as owning her resources"; and "If dialogue is to work, then resource control is highly needed by the people of the Niger Delta."

The analysis of the data from Faith Baptist on this yielded two broad categories: approval and disapproval of violence seen from consequential and deontological perspectives. In ethics consequential arguments assess the rightness or wrongness of an act or behaviour on the basis of its outcome, whether positive or negative. From this perspective, according to Robin Gill, murder, for example, would be seen as being wrong not of itself but because of its possible consequences. On the other hand, deontic ethics evaluates the propriety or otherwise of an act on the basis of a norm or measuring standard (Gill 1985, 5-7). Under 'approval' are participatory approval and non-participatory approval. The non-approving responses fall under deontic disapproval and consequential disapproval. The details are as follows:

### *7.7.1 Approval of Violence*

There were four members of the church who endorsed the violence being perpetrated by some militant groups in the region. The first member who is a university student even acknowledged that he had personally participated in a form of the violence. Thus he says, "Members of this church have not been involved in the shooting or in the killings but I've personally been involved in demonstration and in the beating up of some people who are within my area of operation, especially at home and in Port Harcourt, and I'm still willing to do more." The above is an example of participatory approval. Although he acknowledges that no other member of the church is involved in the militancy, he testifies of his personal involvement. It is noteworthy that there were spontaneous screams of disapproval from the rest of the members present at the focus group session where this statement was made.

Second, there is non-participatory approval, which refers to those who argue that the violence is justified because it is the last resort without being personally involved. A young man who runs his own small-scale business hinged his support for violence on the neglect, marginalisation, and lack of amenities in the local communities, coupled with political awareness; hence he declares as follows:

> The problem with the Niger Delta is neglect by the oil companies, by the Federal Government and by the state governments. People are marginalised; they don't have amenities, even though there is electricity everywhere, but in the communities there is no electricity. So everybody wants to have it by force because there is political awareness now than ever before. Everybody wants to say, 'look, we have these things and God has blessed us with them, why is it that we are not having it for our social and economic wellbeing? So, for that reason, we must fight and the fight will continue.

A secondary school student in the church underscores this perception that violence is the last resort when he argues as follows:

> I think it seems as if violence is the last resort for Niger Delta people because if you try dialogue with the Government, they would just pay the chiefs off. In a situation like, the boys have no choice than to fight for what they think is right. They've found out that since the Government is in control and has paid off their uncles and fathers who are supposed to settle

these issues, they feel that what is left is to take up arms than to wait in the house to die.

### *7.7.2 Disapproval of Violence*

Apart from the above approvals of violence in the church, there was also a strident voice of disapproval. This disapproval could be subdivided into two subcategories: deontic disapproval and consequential disapproval. An adult male member argues that since violence never solves any problem, it should outrightly be rejected and dialogue adopted in its place. In the context of the Niger Delta he feels that the Government should take the initiative in the dialogue and adds that the best remedy against the whole problem is good governance. The following are his comments:

> I like to say that violence solves no problem, whether in Nigeria or South Africa; so there must be a modern means and the only way is to sit at a table and talk. Two countries (may be) at war; while the war is going on, they may sit down to engage in dialogue. That's what the Government should be doing. So if people think that violence will solve any problem, it never does. And the issue of marginalisation in Nigeria, some particular groups tend to cry more than others. I'm from a minority, I'm from a highly marginalised area: the roads in my area have been the same since I was a youth but I'm not sure that young boys from my area will go on rampage. So, violence solves no problem. The onus of dialogue is on the Government and on those who sponsor or use these boys. I've always said that the remedy against a coup is good governance but, as long as we have bad governance in the country, these things would continue to happen.

There were also those who rejected violence on the grounds of Scripture. A young lady in the church who works in the state judiciary spoke in this light when she says:

> Hebrews 12:14 states that we should 'pursue peace with all people and holiness without which no one will see the Lord', and v. 15, 'looking carefully lest any one fall short of the grace of God; lest any root of bitterness springing up cause trouble and by this many become defiled'... Jesus Christ does not have interest in violence, and that is why on the day he was arrested, when Peter brought forth his sword to cut off Marcus' ear, he wasn't happy about it.

Two points are made by the lady above, namely, that believers are called to pursue peace and that violence is contrary to the teachings of the New Testament and the pattern of life seen in Christ.

Moreover, there were those who adopted a consequential approach in disapproving it. For example, a boy of about sixteen years rejects violence because, according to him, "if we use violence in our state or Niger Delta, innocent people would be killed and when they are killed, we'll not have people in the environment again." Thus, his major ground of disapproval of violence in the region is the loss of innocent lives and the problem of depopulation that goes with it.

If violence is not the right way of expressing dissent to the perceived injustice meted out to the region as argued above, then what is the alternative approach? The unanimous answer from the members is dialogue. This is represented by the following statement by a mother in the church who is a civil servant: "I agree that violence is not the option to solving the problem. Dialogue is the only way." But if dialogue is to work, some members highlight some issues that must be addressed. For instance, while agreeing with the suggestion of dialogue as preferable to violence, a mother posits in the following words that it must be based on an 'open-door policy': "Yes, I agree in dialogue, I also want to add that there should be an open-door policy." An elderly male member sees dialogue as a 'hypothesis' that has not worked in the past; thus he says,

> From what they are saying, I want to say something. When you say dialogue is the solution, to me it looks like hypothesis. When you talk of dialogue, dialogue has been held several times. We were in Abuja for several months. Youths have been called and even supervised by Obasanjo himself. So dialogue is a solution but to us it looks like mere hypothesis. So what is there to me is to listen. We have drifted from resource control to revenue sharing. We are only talking of 25% and somebody is insisting that unless we accept 17% he won't agree. Why not try that 25%? What of these youths who have been graduating from universities and there are no jobs for them? You force them into national service and yet you don't pay them their allowances. Also pensioners are not paid.

Both the endorsement and condemnation of violence in Faith Baptist cut across gender and age on various grounds, with the majority rejecting it. Significantly, when the view of a group of twenty nine boys aged between ten and twenty five in the church was sought on whether they would personally wish to get involved in the violence by joining any of the gangs, all of them said they would not, under any circumstance, because they are Christians. Their position was that despite the perceived injustice against the region, violence is not the way out. Thus, apart from the desire for personal safety and communal-preservation, the key basis of preference for non-violent approaches in resolving the problem as seen by majority of members of Faith Baptist is in the person, teaching and ministry of Jesus Christ. In First Baptist the support for violence came mainly from the male folks; while there was zero approval of violence in Hilltop. The support and disapproval of violence in First Baptist were both based on certain teachings of Scripture. On what basis did members of Hilltop reject violence? There was no reference to any bible passage; neither was there any to any Christian ideal. Rather, they rejected it on the ground of its demerits as they claim that it is anti-development. In summary, the clarion call of most members of the three churches was for dialogue. But they insist that if it is to work, the problems of unemployment, illiteracy and poverty must be addressed.

## 7.8 Marginalisation, Oloibiri and Resource Control

These three themes have already been foreshadowed in the previous sections of this chapter and due to their interrelatedness, are treated together. As indicated in the chart at the beginning of the chapter, the themes of resource control, marginalisation and Oloibiri respectively featured a total of twenty three, sixteen and ten times in the three churches. The logical flow of these three themes as deduced from the data is the belief that there is marginalisation in Nigeria, which, in the case of the Niger Delta, is best symbolised by Oloibiri, the first community from where oil was drilled in Nigeria; and the remedy to the marginalisation is resource control. Some of the expressions used to convey the idea of marginalisation include 'oppressed', 'neglected', 'discriminated against', and 'colonised'. The discussion that follows will illustrate this.

Reference was made earlier to a female lecturer in Faith Baptist who attributed the root-cause of the conflict to the marginalisation of the people of the region because of their minority status, despite the fact that they produce the bulk of the nation's revenue and foreign exchange. The same sentiment was expressed by a youth in Hilltop, who says:

> The people of the Niger Delta are being colonised in the region because we are not in the majority and the majority were oppressing us and we didn't have much say. Now, the issue of revenue derivation as it is within the Nigerian context is lopsided because we do not have what we ought to receive; yet we are the people suffering all the consequences of oil exploration and exploitation in the Niger Delta. Also, I want to submit that the Federation we are running now is lopsided and needs to be addressed.

What is the evidence of the marginalisation? A number of answers were given. For instance, a youth in Hilltop categorically states, "The issue borders on having taken over US$60 billion from a place and neglect of the people, marginalisation and environmental degradation." A retiree from the nation's publicly owned oil company, NNPC, adds emphatically from his personal experience, "There is discrimination against the Niger Delta indigenes in certain level of management in the oil industry. For example, I was prematurely retired from NNPC in 1996 along with three other ... Deputy Managers; whereas some people from other parts of the country were employed to replace us." Thus, the major reasons for claims of marginalisation against the region that were adduced in the three churches include neglect, lack of political voice, lack of employment opportunities and discrimination against Niger Delta indigenes in management positions in the oil industry, environmental degradation, and imbalance in revenue allocation.

Who are the perpetrators of the marginalisation? At least three answers were noted on this question. The first response is that it is being carried out by the three majority ethnic groups. This could be deciphered from the following contribution by a lawyer in Faith Baptist:

> We just finished from our NBA (Nigerian Bar Association) conference and we canvassed the issue and thank God that it was an Igbo man that delivered the lecture and said that if this oil were to be in the north (among) the Hausa-Fulani, that law would not stand, they would have amended it. If it were to be in the West, in Yoruba land, they would have amended it; if it were to be in his native Igbo land, it would not stand, they would have amended it. But because it is in the Niger Delta the law is there and there is nothing we can do.

There are those who see the marginalisation from a regional perspective and therefore argue that it is the north that is oppressing the south. An adult male in Hilltop conveys this notion when he says, "The truth of the matter is that Niger Delta people are really suffering and the people up, which are the four cardinal points we have – the north, the south, etc. – the north is pressing the south. So if the north gives justice to the south then there would be peace."

However, there are those in the three churches who see a third layer of the discrimination. They claim that this layer operates within the region. An example is a middle-aged man in Faith Baptist who hails from Delta State. He says, "On the issue of marginalisation in Nigeria, some particular groups tend to cry more than others. I'm from a minority within my own state; I'm from a highly marginalised area. The roads in my area have been the same since I was a youth. But I'm not sure that young boys from my area will go on rampage." Thus, marginalisation or discrimination is not a one-sided phenomenon.

A recurring theme in the midst of all the discussion in the three congregations and which seems to be the greatest symbol of marginalisation against the region from their viewpoint is Oloibiri. Oloibiri is in present day Bayelsa State and was the first place oil was struck in commercial quantity by Shell-BP in 1956. This is illustrated by the following statements by some members of the three churches. A youth in Hilltop states as follows:

> I have to submit that crude oil in the Niger Delta is a curse to us. I want to state that the first oil well was struck in *Oloibiri* in Bayelsa State in the year 1956, but anybody who cares to know should go to *Oloibiri*, as I talk now, and see whether the oil that was first struck there is a blessing to the people or a curse... I want to state that it is a curse. If you go to Bomu in Gokana, or Dere, where oil was struck in 1958 the people are still battling with Shell.

The following contribution by a member of First Baptist is in the same line of thought: "I want us to refer to two basic things. First, oil was found in *Oloibiri* in present-day Bayelsa, but today, how is *Oloibiri*? Nothing grows there; there is no tangible activity that is vibrant in the place due to oil exploration." Then two members of Faith Baptist, a senior civil servant and an engineer, respectively, add as follows:

> Well, the Niger Delta question is a serious one indeed. The point had been made earlier that oil was discovered at *Oloibiri* in 1956. But it's sad to note that there is no road to *Oloibiri* till today; there is no school, no hospital, and people live in squalor, in poverty, in penury, name it. I know that if *Oloibiri* were to be in some other part of this country, it would have been a tourist centre, a centre of attraction. But today the reverse is the case. It's in darkness.

And

> The Niger Delta people, looking at the picture painted by *Oloibiri*, have realised that oil is not a perpetual resource; it will finish. Recently, they said that Shell wants to go back to *Oloibiri*, but the oil found in the place finished and the place was left the way it is today. One day this oil is going to finish and they'd have taken all that came with it out of the Niger Delta. Then that way we'll be left to die the way we are. Oil is not a perpetual mineral; so we should be allowed to use it and develop our area.

The points on Oloibiri from the foregoing could be summarised as follows: First, for a number of the members of the three churches, it perfectly illustrates the curse of oil as its environment has seriously been degraded as a result of oil exploration activities. Second, despite the fact that oil was produced from there, presently it lacks basic social amenities such as schools, hospitals, roads and electricity, and the local community lives in abject poverty. Third, it aptly portrays the exhaustible and transient nature of oil deposit. And fourth, in order to avert a repeat of the Oloibiri case, they argue that the people should be allowed to use the revenue from oil to develop their area. This is expressed in the form of a call for 'resource control'.

A youth in Hilltop defines resource control as "the people want to control their resources"; while another emphatically argues, "*resource control* is highly needed by the people of the Niger Delta." On her part, a middle-aged mother in First Baptist hinges the return of peace on it; hence, she says, "For us to get peace, equity in this our land, the region must be given her right, right as *owning her resources*." It is pertinent to note that in each of the focus group sessions where this topic was discussed, the members who spoke did so with a depth of passion and emphasis and were overwhelmingly supported by the majority of the rest of the members of the churches present. As in the discussion on violence and non-violence, a few supported the enforcement of resource control through violence, while the majority preferred non-violent approaches in bringing it about.

## 7.9 Summary

This chapter examined the socio-political consciousness of the three congregations regarding oil and conflict so as to ascertain whether or not they are indifferent to their social setting and/or maintain a fatalistic attitude towards the conflict. Evidence from their experiences, statements, actions and ministry within the context of the crisis is at variance with the claims. Rather, they manifest a good grasp of the situation as seen from their analysis when set against the country's political economy of oil and conflict described in Chapter Three from primary and secondary sources.

# CHAPTER EIGHT

# CONCLUSION

## 8.1 Evaluation

This study set out to examine Christian theological and socio-political consciousness in the Niger Delta region of Nigeria within the context of oil and conflict from Baptist and Pentecostal perspectives and to use the data gathered from three local congregations in the region to assess the predominant functionalist and instrumentalist reading of African evangelical and 'fundamentalist' Christianity regarding social responsibility as best represented by Paul Gifford. These were specifically addressed in the preceding chapters.

To sum up, it is now pertinent to address the key specific claims of Gifford that encapsulate the reading. These were highlighted in chapter two and include: (1) that such Christianity lacks social responsibility and is apolitical, since it is indifferent to the plight of the wider society; (2) that it encourages fatalistic acceptance of disasters and misfortunes; (3) that it is anti-development; (4) that it places emphasis on unwavering loyalty to government, the evil of the world, spiritual warfare between good and evil spirits, dualism of body and soul, and heavenly reward, since, according to him, "nothing this-worldly matters" to them; and (5) that its overriding preoccupation is evangelism.

On the first claim, evidence from the research is to the contrary as the three churches in the study demonstrated a high degree of social awareness and responsibility as could be deduced from a comparison of their analysis of the crisis with the political economy of oil and conflict described in chapter three. They manifest genuine concern for the condition of their social setting by their prayers, statements, and social witness. If, for instance, the plight of the environment does not concern them, then Hilltop would not engage in road maintenance or free rural health programmes. On the second claim, there was no evidence of acceptance of the conflict in the region as the will of God or as inevitable. Rather, they show significant disgust and anger at the crisis and its perpetrators. This is evident from their grasp and analysis of the problems and challenges posed by crude oil and conflict as demonstrated in chapter seven.

Throughout the period of research and in the course of analysis of the data, this writer did not see any indication of opposition to development as the third claim above says. On the group of assertions in claim No. 4, the study did not see proof of emphasis on uncritical loyalty to government or of dualism of body and soul. Rather, there was evidence of belief in evil forces and therefore spiritual contest between good and evil spirits and in heavenly reward. Indeed they believe that there are evil powers principally represented by the devil and demons, as well as evil forces, behind such misfortunes as miscarriage, witchcraft, and childlessness. Nonetheless, stemming from the belief that Jesus is the ultimate *Oga Kpatakpata* and Bulldozer, they also affirm that he is the Conqueror and vanquisher of Satan and his agents, as well as his agencies, and that the believer is called into the ministry of enforcing that victory. Apart from the biblical tilt to this belief, it is also vitally connected to the African worldview and is not peculiar to African evangelical Christians. This point has been noted by a number of scholars of African Christianity (See Walls 1996, 5-11; Sindima 1994, 134-138; Kalu 2008, 175-186; and Dyrness 1990, 42-52). For example, John Parratt remarks, “It needs scarcely to be emphasized that the traditional world was, like that of the Gospels, one in which supernatural powers impinged on the world of humans at all points, and in which all life was subject to the influence of spiritual powers” (Parratt 1995, 59; cf. Mana 2004, 95-96).

Moreover, in as much as they emphasise heavenly reward, they do not lose sight of their earthly existence, condition and responsibility. Hence, they pray and fast for betterment in this life in the form of breakthrough, prosperity, and health. A person that is so heavenly minded that he/she lacks this-worldly focus would not be praying and fasting as hard as these people do for better opportunities in this life. Their spirituality is a form of political praxis in which they seek to “shape history through prayer and fasting” (Kalu 2008, 196), or what Kalu describes as “intercession as political praxis” (Kalu 2008, 219).

In addition, Gifford’s analysis fails to take cognizance of the social skills these churches impart to their members through their internal organisation in which members “learn how to function democratically, elect their own officers, develop and exercise leadership skills, handle money and budgets, plan and execute projects, and critique the results” (Kalu 2008, 196). The following words by a member of Faith Baptist, who once served as the Church Secretary and who now runs a thriving private business, illustrate this:

> One of the things I thank God for, for the opportunity I had to serve as our Church Secretary, is that the things I learned from the experience are helping me to run my fishery business now. I learned how to set goals and plan and follow through the plans; and how to relate with staff, as well as time management. Without that opportunity, I'm not sure I would have known these things.

Among all the claims by Gifford, perhaps the one that is closest to the truth is the fifth one, namely, that evangelical Christians in Africa place much emphasis on evangelism. This was found to be the case in each of the three congregations. However, the error is that whereas soul winning is their *primary* focus, it is not their *sole* preoccupation. Moreover, from the investigation, out of the six dimensions of conversion outlined in chapter six, the ones that feature in Gifford's analysis are the *ontological*, *existential* and *teleological*, thereby leaving out the *moral/behavioural*, the *political* and the *spiritual*, which are at the heart of their understanding of ecclesial social role. Drawing from their understanding of Christian conversion as a multi-dimensional experience, they teach, preach and pray for peace and justice and majority of them espouse non-violent modes of agitation against the perceived injustice in the region. Hence to them, the power of conversion gives *freedom*, *life*, *justice*, *transformation*, *control* and *hope* to the believer. These core values of Christian conversion from their perspective together form the nucleus of their attitude towards the wider society. As described elsewhere by this writer,

> It is an attitude with an inside-out orientation that places primary emphasis on changing the individual before changing the society, and which sees the above six values as the privileges of those who are converted and who then should seek to expand the network of the beneficiaries through soul winning. It does not mean that such churches are anti-development or socially indifferent (Osuigwe 2010, 300-301).

Also, it is an attitude that sees the primary responsibility of the Church to the wider world as that of being an exemplary community that mirrors the life and teachings of Christ by "deliberately positioning itself as a response to what are represented as corrupt or ruined religious and political traditions" (Marshall 2009, 2-3). This perspective to Christian socio-political involvement has also been espoused by a number of leading theologians such as John Howard Yoder (Yoder 1972; 1994), Stanley Hauerwas (Hauerwas 1986; and Hauerwas and Willimon 1992), and Emmanuel Katongole. For example, Katongole writes that the Church is a community of "people formed by the story of Christ's life, ministry, death, and resurrection as contained in the scriptures and witnessed within various historical communities across time" (Katongole 2002, 141).

Therefore, having reviewed all the evidence from the field research, it is appropriate to conclude that the claims that African evangelical and 'fundamentalist' Christianity lacks social responsibility and is apolitical and anti-development are not applicable to First Baptist Church, Buguma City, Faith Baptist Church, Port Harcourt, and the International Christian Centre, Port Harcourt.

A pertinent question that arises is, Why this discrepancy between most of these claims and the reality in the three churches? A possible answer is the difference in the location of these congregations – they are based in Nigeria – whereas the dominant spheres of Gifford's analysis are Ghana, Uganda, Zambia, Cameroon, and Liberia. But this is vitiated by the fact that most of the claims are general and generic in nature. Another answer could be that these churches belong to his 'except' churches within his classification of African evangelical/fundamentalist Christianity for, he says, "There are some fundamentalist churches that do engage in development projects." But he then cancels this out by saying that "their involvement is really a contradiction of their teaching, for their message itself militates against such commitment" (Gifford 1991, 19). This answer, therefore, is not satisfactory because a close reading of the case studies would show that there is no contradiction between their teaching and practice regarding Christian social involvement since they believe that "the earth is the Lord's and the fullness thereof." Richard Burgess puts this into perspective from his study of Igbo Pentecostalism in south-east Nigeria:

> Contrary to Gifford's assertion that much of Africa's Christianity (i.e. the Pentecostal sector) dismisses the contemporary political situation as theologically irrelevant, many Igbo Pentecostals are very conscious of it. While they lack a formalised political theology, it is implicit in their discourse and praxis (Burgess 2008, 281).

In the light of the foregoing, the answer for the discrepancy is located elsewhere. As indicated in chapter two, a number of scholars who have reviewed some of Gifford's publications (for example, Allan Anderson, Ogbu Kalu, Kwame Bediako, and David Maxwell) attribute it to certain factors, which include his methodology, his perspective, and/or his conclusions, some of which they consider inaccurate. The findings of this study confirm these observations. It is apparent that his methodology is heavily dependent on the social sciences and his perspective is dominated by European enlightenment values. These appear to have led him to some conclusions that are at variance with the way these churches conceive their world and their role in it vis-à-vis their faith and the "African maps of the universe" (Walls 1996, 5). These submissions could be substantiated with the following three facts emanating from the research.

The first is Gifford's apparent lack of appreciation of their ecclesiology, particularly their theologies of conversion and prayer, within the spectrum of their Christology and doctrine of God in relation to their conception of the church's role in the wider society. This is because their theology of prayer develops out of their doctrine of God; their ecclesiology is connected to their Christology through their soteriology; while their social ethic is intrinsically linked to their Christology, ecclesiology and soteriology. That is, they believe that the church is composed of voluntary converts, people who are converted and are being transformed by faith and discipleship to the life, ways and teachings of Christ. Second, it is obvious that in his analysis Gifford fails to take seriously their worldview, despite the fact that he acknowledges it, due mainly to the age-old problem of 'insider' versus 'outsider' perspectives. The third reason is Gifford's overtly undue dependence on Western-oriented instrumental and functional analysis of the socio-political role of African Christianity. This is why Kalu warns that "Sociologists of religion may miss the driving force of religious power in religious movements by paying too much attention to functions of such movements in social structures" (Kalu 2008, 248).

Also, Ruth Marshall rightly argues that a prerequisite to understanding Pentecostal explosion in Nigeria and its "complex political productivity in a nonreductionist way" is the "rejection of a priori understandings of the relationship between the religious and political, and hence the extraction of analysis from the opposition between faith and reason that is still central to many forms of critical inquiry today" (Marshall 2009, 3). Continuing she adds, "Studies of the political significance of Christianity in Africa have too often presumed the models of the relation between religion and politics with which social science represents Western society" (Marshall 2009, 19). David Maxwell falls into the same template with the above when he argues that "much of the scholarly debate of born-again religion has been dominated by functionalist or instrumentalist questions about its relations to politics and economics. There is a danger that such approaches miss the point, ignoring the movement's own idioms and intentions" (Maxwell 2002, 176; cf. Sanneh 1983, xi; and Isichei 1995, 8).

However, the above notwithstanding, there are certain points of agreement between the data and some other findings of Gifford in his extensive studies on African Christianity, especially on Pentecostalism, which also tally with the findings of other scholars. For example, the research shows that he is right in his delineation of Pentecostal worship services into three parts – 'praise and worship', offering, and sermon. This was found to be the case in Hilltop. Second, his description of their Bible interpretation and preaching style is apt, as has also been noted by other scholars such as Kalu, Asamoah-Gyadu, Burgess, Ukpong, and van Eck. In the third place, the situation in the two Baptist Churches confirms his assertion on the 'charismatisation' of the missionary founded churches, a phenomenon others, such as Bediako and Asamoah-Gyadu, have also noted and called 'pentecostalisation'.

## 8.2 Total Freedom and Social Transformation: Towards a Political Theology for Nigeria's Oil Economy

It is now pertinent to examine and propose a contextual theology of social and political engagement for the Niger Delta. Two themes borrowed from the case studies that encapsulate the theology of the three congregations within the context of the conflict are "total freedom" and "social transformation." The key categories that convey these themes are Christology, ecclesiology, spirituality, and ecology. These themes and categories will be summed up to propose a contextual political theology for the Niger Delta. The discussion will conclude with a comparison of the proposed theology with liberation theology so as to bring out their respective points of affinity and variance.

### *8.2.1 Total Freedom and Social Transformation*

Reference was earlier made in chapter four to the annual theme of First Baptist for 2006 which was, "My Year of Total Freedom," as well as the church pastor's explanation of the background and basis of the theme. According to him, the city, the state (Rivers State), and the entire region had been under spiritual, economic, social, and political bondage. On how to end this 'bondage', he explained that it is through total freedom made possible by God, "using human instrumentality."

This theme of total freedom permeated all the programmes and activities of the church throughout the year 2006, and is reflected in the other two churches, especially in their prayers, as they besought God for freedom in the areas mentioned. The theme carries a number of subthemes, namely, the possibility of total freedom, the scope of total freedom, the power for total freedom, the timing of total freedom, and the purpose of total freedom. On the possibility of total freedom, the declaration simply assumes it in a number of ramifications, namely, the spiritual, the moral, the religious, the social, the political, and the economic.

A couple of meanings are attached to 'spiritual' freedom. First, it means salvation of the soul; and, second, it refers to freedom from such practices as idolatry and cultism. 'Moral' freedom refers to emancipation from tendencies toward various vices, including sexual promiscuity, and the capacity to live a life of fidelity and chastity; while 'religious' freedom refers to the liberty *to* worship and freedom *from* forceful participation in traditional religious activities and/or rites. The social, political and economic expressions of total freedom are portrayed as freedom from marginalisation. The belief in the churches is that freedom from marginalisation is possible; hence their present state is neither inevitable nor unchangeable.

If total freedom is possible, how does it come about? Or what power can bring it to pass? The answer is nuanced and manifests in a number of trajectories. It is believed that God is the Source of total freedom, in conjunction with "human instrumentality"; Jesus makes it possible; while the church is the arena where it operates in its fullness. In relation to the timing of freedom, the belief is that it is determined by God. However, total freedom is not conceived of as an end in itself but as a means to an end. Temporally it is geared toward social transformation and, eternally, its goal is heaven for the believer. Social transformation begins with the transformation of individuals, following their conversion, before extending to the society. The means of this transformation include their teachings (*orthodoxy*), practices (*orthopraxis*), and experiences (*orthopathos*), such as sermons, Bible studies, pastoral activities and programmes, worship, prayer, and songs, which together form their "political spiritualities" (Land 1994, 13; Marshall 2009). A discussion of their spirituality and political ecclesiology will follow shortly.

### *8.2.2 Oga-Kpatakpata as Saviour: Jesus' Way of Justice and Peace*

The case studies showed that the three congregations believe that Jesus is the ultimate *Oga-Kpatakpata* who stands for peace and justice. The aim here, therefore, is to elaborate on that social ethic and how it applies to the conflict. Several opinions have been expressed over the last century on the socio-political implications of the image and message of Jesus. At both ends of such views are two extreme positions: the image of Jesus as a preacher of purely a-political moral and eschatological message and the image of Jesus as a full-blown political agitator at the head of a failed revolution (Ukpong 1995, 1 and 10; see footnote numbers 1, 2, and 3 on p. 10). But there is a growing consensus that this dichotomy is not necessary as neither extreme portrays the whole picture (Powell 1998, 174). Rather, evidence from the NT suggests that Jesus indeed refused to be seen as a political messiah: during his temptation in the wilderness (Matt. 4:1-11; Luke 4:1-13); after his feeding of the multitude when the people wanted to make him king (John 6:14-15); and his refusal to adjudicate in the dispute over inheritance by two brothers (Luke 12:14). Yet his disciples persisted in their erroneous belief that he was a political messiah: cf. Act 1:6 and Matt. 8:27-33 (Ukpong 1995, 1-2). But Jesus neither sought nor encouraged his disciples to seek political authority. This does not mean that he did not say or do anything in critique of political authority and the politics of his day. In other words, Jesus was not a-political, but was a threat to the powers that be. How?

Jesus was a political threat simply by his preaching of the coming of the kingdom of God that he came to inaugurate. Such a kingdom is one of justice, peace, tolerance and nonviolence. It requires that God's will be done on earth just as it is done in heaven (Matt. 6:9-13). The message was good news to the poor as it meant the shattering of the economic structures that impoverished them and the pulling down of the political system that ensured and enforced the exploitation. Also, Jesus preached a message of liberation of the oppressed and marginalised in society (Ukkpong 1995, 3). At the same time, it was a message that attacked the oppressors, both political and religious, and those who wanted a continuation of the status quo (Ukpong 1995, 3).

Specifically, how did Jesus respond to the unjust and oppressive system of his time, especially in the light of the prevalent perspectives among various Jewish groups? According to John Howard Yoder, he was faced with four options, or "Four Ways," as seen in the various responses of the Jewish people/sects to Roman occupation: *realism* – accept the situation as it is and hope for the better – which in Jesus' time is represented by the Herodians and Sadducees; *"righteous revolutionary violence,"* represented by the Zealots; *withdrawal* to the desert so as to escape the pernicious corrupting influence of the society and the system on which it is built, as seen in the Essenes; and the option of "*proper religion*," which is seen in the Pharisees who sought to keep themselves pure and untainted by adhering to rules of segregation and a posture of non-involvement (Yoder 1972, 18-27; 2003, 30 and 57; cf. Bediako 2004, 103-104; and Richardson 1973, 12-23).

But, according to Yoder, Jesus did not choose any of these options. Rather, just as God called Abraham or Moses or Gideon or Samuel, Jesus "gathered His people around His word and His will" and thereby created a society unlike any other human society and outlined for it seven "Political Axioms" in the Sermon on the Mount, including an ethic of reconciliation (Matt. 5:22, AV). The other ethics mentioned by Yoder are the *ethic of repentance* (Matt. 4:17, 23-35), *ethic of discipleship*, (Matt. 5:1-7), *ethic of testimony* (Matt. 5:13-16), *ethic of fulfillment* (Matt. 5:21ff.), *ethic of perfect love* (Matt. 5:44-48), and *ethic of excess* (Matt. 5:47, KJV). From Yoder's analysis, the society Jesus formed was a voluntary one in which people could belong to only by repenting and freely pledging loyalty to its King; it was a society with mixed composition. It was mixed racially, religiously and economically; and it was a society in which Jesus gave its members a new way of life grounded on love (Yoder 1972, 29).

Deriving from the above, the church in the Niger Delta should avoid the 'Four Ways' in Nigeria's political economy of oil and conflict outlined in chapter three, namely, the way of the Federal Government and the MNOCs, the way of state and local leaders in the region, the way of civil society groups, and the way of militant youth groups as none of them conforms to the way of Christ.

### *8.2.3 Oga-Kpatakpata as Bulldozer: Jesus and Power*

A major finding of this research is that there are at least five discernable lines of claims of marginalisation in Nigeria: international capital marginalising the country through the MNOCs; the north marginalising the south; the majority ethnic groups marginalising the minority groups; the elites marginalising the "common masses"; and minorities marginalising minorities. A sixth one that could be added, but which was not mentioned in the case studies, is the marginalisation of 'non-indigenes', that is, Nigerians who live in States or Local Government Areas other than the ones they have their ancestry (See Human Rights Watch 2006).

The bedrock of marginalisation in Nigeria is the abuse of economic and political power, and this culture of power abuse predates colonialism but was strengthened by it (Bediako 1998, 3). Thus, political and economic leaders see power as a means of self-promotion, self-enrichment, group extension, and the promotion of ethnic, religious and/or regional interests. As a result, the various grids of marginalisation exist whereby those not favoured by a particular grid and who are outside its circle of beneficiaries cry wolf, while at the same time participating in, encouraging, abetting and/or championing another grid. Hence, in this system of *diffused marginalisation*, no one has the monopoly of the practice. The beneficiaries or victims of each grid are determined to a large extent by those who control or have access to power. Put differently, the *oga-kpatakpata* of each setting determines who is in and who is out. Consequently, it is imperative to consider the theological meaning of power as demonstrated by Jesus.

From a theological viewpoint, Jesus epitomises the true meaning of power. Again, with reference to the cross, through his life, death and resurrection Jesus demonstrates that true power comes by way of self-donation and seeming powerlessness and the fact that its greatest purpose is humble service rather than domination. This understanding does certain things that touch on the theological definition of power. First, it shows that Jesus de-sacralises power through the cross by inverting its meaning from domination to sacrificial service. Kwame Bediako has noted that the "way of Jesus" is the way of "non-dominating power" and adds, "The mind of Jesus on the questions of politics and power is not a dominating mind, not a self-pleasing or self-asserting mind, but a saving mind, a redemptive mind, a servant mind ... Jesus' way of dealing with political power represents the perfect desacralisation of all worldly power" (Bediako 2004, 101). Second, through the cross Jesus demonstrates the inherent power in seeming powerlessness as it prepared the way for his ultimate victory over death through his resurrection and subsequent exaltation (Bourne 2007, 197-198).

Third, not only did Jesus gain power through seeming powerlessness, he exemplifies the concept of shared and participatory power, which is power permeated with love. While walking/working with his disciples, he shared power with them in the various assignments he gave them. Again, after his resurrection, and prior to his ascension, he instructed his disciples to remain in Jerusalem until they had received power and, when the promised power dawned, it came upon all the believers as recorded in Acts 2. Ultimately the purpose of the power was for kingdom service, namely, to be his witnesses, and not for self-aggrandisement. Thus, the Church was established on the basic principle of shared power as every member is endued with some power from the Holy Spirit in the form of spiritual gifts. From a Christological perspective, therefore, power is packaged in humility and delivered through service and its true test is its effectiveness and service, not domination and control (Gifford 1998, 17). This is the way the church should be and what it should insist on and proclaim in the Niger Delta.

### 8.2.4 Embodied Community as Political Praxis

It was indicated earlier that the church is the arena for the experience of total freedom. This section will now elaborate on that by describing a political ecclesiology for the Niger Delta from the case studies and thereby tease out David Martin's assertion that "evangelicals and Pentecostals are political because they try to alter what is in their capacity to alter, beginning with themselves, their families and congregations" (Martin 1996, 12). The aim is to demonstrate that the church is a form of political praxis. The idea of embodiment, which is defined as "an existential condition in which the body is the subjective source or intersubjective ground of experience" (Csordas 1999, 143), is made prominent by three phrases used in chapter six: 'body of Christ', 'representative of Christ', and 'ambassadors for Christ'. Underlying their usage is the belief among members of the three congregations that there is an existential connection between Christ and the Church in which the latter represents the former as his 'ambassador'. In line with this embodiment motif, the Church in the Niger Delta as the body of Christ ought to incarnate him by projecting the new humanity revealed in and congruent with its *Oga-Kpatakpata* in character, conduct, and commission. This section will examine the ontology and purpose of embodiment and then indicate their implications for the Church.

Borrowing terms from the field of anthropology of religion, namely, self-awareness, self-preservation, and self-ownership (See Csordas 2002, 58; and Tierney 1999, 233-234), the ontology of ecclesial embodiment in the context of the Niger Delta could be subdivided into collective self-awareness, collective self-preservation, and collective self-ownership, with four additional sub-categories – collective self-governance, collective self-propagation, collective self-transformation and collective self-donation. Collective self-awareness refers to the way the three churches view themselves; collective self-ownership is how they view the ownership of the Church; collective self-preservation is the modes of actions they take or seek to take for their sustenance and protection within the context of conflict; and collective self-governance refers to how they govern themselves. These four aspects relate primarily to their ontology of embodiment, while collective self-propagation, collective self-transformation and collective self-donation are tied to their purpose of embodiment and are taken up below.

On their collective self-awareness, each of the three congregations defines itself as a body or congregation of believers that meets in a particular place for worship, teaching, fellowship, ministry, prayer, and outreach. This is intrinsically linked to their collective self-ownership in which they see themselves as belonging to Christ and to one another. Within the context of the conflict, how have they sought to ensure their collective self-preservation? Their main modes of action are basically the same: pastoral and material assistance and prayer support. The first point of call of members when they face the dangers of the conflict is their respective churches and, when they do so, their churches respond by way of provision of temporary shelter, food, and prayer.

An area in which the two sets of churches (the Baptist and Hilltop) show a paradox of similarity and dissimilarity is their collective self-governance. They are participatory in orientation and claim Jesus as their ultimate authority. Where they differ is on the outworking of that authority among their members. Whereas the two Baptist Churches operate under democratic principles in which power resides with the congregation, Hilltop practices participatory hierarchy in which power is vested in the hand of the leadership in a mixture of episcopal and presbyterian styles of leadership.

What is the purpose of ecclesiological embodiment outlined above? The answer is provided next. There are two related questions pertinent to the discussion that will be addressed, namely, whether obedience to the Great Commission is solely for the purpose of soul winning, and whether there is a link between spiritual conversion and social liberation. These two questions have antecedents in Christian theology. For instance, Jesse Mugambi has demonstrated that Karl Barth represents the 'salvationist' view of the purpose of the church with focus on personal salvation; while Soren Kierkegaard and Ludwig Feuerbach exemplify the position that the church should be involved in the affairs of the world, i.e., the liberationist view, which emphasises social transformation (Mugambi 1995, 6). From the perspective of the case studies, the answer to the question of the purpose of ecclesial embodiment could be coalesced into three schemes in which the first is the spiritual and pastoral, the second the moral and the third the socio-political. They are respectively collective self-propagation, collective self-transformation, and collective self-donation. Collective self-propagation refers to the way they go about what they consider to be their primary task, namely, evangelism; collective self-transformation focuses on how they see their role towards those within their fold in relation to Christian formation, and this is done through teaching and discipleship training; while collective self-donation is how they view and what they do or seek to do regarding the Church's socio-political role.

Both collective self-propagation and self-transformation are primarily aimed at individuals and the underlying philosophy is that transformed individuals produce transformed societies. This is the key to the answer to the second issue above regarding the link between spiritual conversion and social liberation. To them the link is in personal transformation. The answer to the first question, namely, whether obedience to the Great Commission is solely for the purpose of soul winning, is now presented. From the preceding discussion on collective self-propagation and collective self-transformation it is apparent that the answer is negative. Jesus' command to his disciples to preach the Gospel to all nations was not intended to be only for statistics – the number of people who accept Christ – but also was for transforming the society through true disciples who model their lives after him. These two aspects roughly correspond to the 'quantitative' and 'qualitative' dimensions of salvation in Liberation Theology (See Gutiérrez 150-151).

Collective self-donation or ecclesial embodiment in the socio-political realm is subdivided into two categories: those that touch on the traditional social sectors, such as education and health, and those that deal with direct political involvement, which some churches in a number of African countries have engaged in (See Gifford 1995, 3). Reference was made in Chapter Six to a mother in First Baptist who identified some social roles of the church such as engaging in philanthropic activities and establishment of educational endowment for the less privileged.. According to her, the Church is not solely for the evangelisation of lost souls, but also for engagement in acts of philanthropy.

On the political arena and with particular reference to the Niger Delta crisis, the three congregations believe that since the Church stands for peace, it should pray for, preach and insist on equity, fairness and justice. In addition, it should take some practical steps in helping to resolve the problem such as mediation, organisation of seminars, workshops, press conferences, rallies and peaceful protests, as well as the issuing of memos and communiqués. Also, churches should engage in leadership development and encourage their members to actively participate in politics, so as to bring positive changes to the system from within. The preceding discussion on the ontology and purpose of ecclesial embodiment has some implications for the Niger Delta conflict.

The first is that the imagery legitimises the body as a genuine theological theme and nullifies the spirit-matter dichotomy. The body is a creation of God which has to be preserved and protected from undue violence since "The first right of natural life is the protection of bodily life from arbitrary killing" (Bonhoeffer 2005, 189). Therefore, the maintenance of the health and integrity of the body is a legitimate theological issue. Churches cannot afford to sustain the ceding of "all claims to the bodies of their members to the civil authorities" (Harvey 2003, 96-97).

The imagery of the Church as body also implies the existence of other bodies and the right of 'the other' to life, fecundity and happiness. Such non-church bodies are the target of its self-propagation efforts and such efforts cannot be effectively done in a disordered society. John Howard Yoder aptly captures this when he says that the "function of the state in maintaining an ordered society is ... a part of the divine plan for the evangelization of the world" (Yoder 2002, 13). Since God is the Creator and Owner of life and *all things*, he is the creator of other bodies, the same bodies the church seeks to evangelise. There are individual bodies, social bodies and political bodies. Each of these bodies is wounded and bleeding in the context of the Niger Delta conflict as a result of violence. Yet, each of these bodies has a right to health and wellbeing within an environment of peace and justice, which the church ought to teach and insist on. A dead body cannot be a target of the church's propagation ministry, since "the dead cannot praise the Lord."

Also, the church as the body of Christ should be the mirror through which the world views Christ. This means that its valuation and analysis of the world should be from the lens of its Lord or *Oga-kpatakpata*. For example, in a context of conflict such as the Niger Delta, the Church's analysis should go beyond the visible, or direct, violence to include identification of its invisible dimensions or the structural and cultural roots of the crisis with a view to addressing them. A major factor responsible for the prevalent culture of violence in the region, for instance, is marginalisation, whether real or imagined. The church can play a role in overcoming this by insisting on and practising a culture of fairness and equity for all and by identifying with the marginalised.

A major ecclesial role in the context of the Niger Delta conflict, therefore, is conflict transformation, which is a process that seeks to address the broader social and political root-causes or sources of a conflict and endeavours to transform the negative energy exerted in the violence into positive socio-political use (See Fisher and others 2000, 7-8; and Jeong 1999, 389-400). Theologically conflict transformation is at the heart of the Christian message of the cross. For, through the cross, Jesus transformed humanity's alienation from God into friendship with him through reconciliation (See 2 Corinthians 5:18-21). Also "the dividing wall of hostility" among humans along the lines of race, ethnicity, social status, and gender has been pulled down in Christ (e.g., see Galatians 3:28 and Ephesians 2:14-18). The Church is in a position, therefore, to apply this to the socio-political realm by bringing a change in focus from mere conflict management or resolution to transformation whereby the energies and resources expended on the conflict are turned into positive use.

Specifically, the Church as embodied community can play the following additional roles within the context of Nigeria's political economy of oil: social – by engaging in the rehabilitation and reintegrating of former militants and by equipping the people with social skills, such as dialogue, thrift, and discipline; economic – by providing microcredit facilities to the poor so as to identify with and empower them and thereby reduce the rate of poverty in the region; and political – by lobbying political leaders for the abolition of unjust laws as well as for reconciliation, justice and peace (Ward 1984, 205). Related to the above is the need for Christian churches in Nigeria to consider the formation of an institutional framework for in-depth theological analysis of Nigeria's political challenges so as to inform and enrich Christian participation in the quest for justice and peace in the country. An example of such a framework is the Christian Institute in Johannesburg, South Africa, which was formed by Christian churches in August 1963 during the Apartheid era. According to Adrian Hastings, the Institute was "an ecumenical group of Christians concerned to bring about reconciliation within South Africa and to relate the gospel to daily experience." Continuing, he observes that in subsequent years it became "by far the most powerful and effective Christian witness to justice in the country" (Hastings 1979, 146). Such a structure is at present lacking in Nigeria regarding the perceived injustice in the Niger Delta. This vacuum could be filled if the churches work together.

In addition to the above, the Church has a prophetic responsibility within the context of the conflict. Prophetically, the churches are faced with three options: (1) to speak out and expect to be criticized by rulers for doing so; (2) to publicly keep quiet – but seek to change government policy by persuasion behind the scenes; or (3) to maintain cordial relations with the government and cooperate with it so as to maintain social and political stability (Haynes 2004, 68-69). Respectively these options could be tagged open-door prophetic witness, back-door prophetic persuasion, and closed-door patrimonial participation. From the perspective of the theology so far outlined, it is apparent that the most credible option for the church is the open-door prophetic witness, even if it draws "vitriolic reactions from the wielders of secular authorities" (Odion-Akhaine 2006, 760). This could then be complemented, where appropriate, with back-door prophetic persuasion. Fortunately the form of prophetic witness espoused here has antecedents in the Barmen Declaration proclaimed by the Confessional Synod of the German Evangelical Christian Church in

May 1934, and the *Kairos* Document produced by a group of South African theologians against apartheid in 1985. The Barmen Declaration, fully known as "The Theological Declaration of Barmen," was written by a group of church leaders in Germany, led by Karl Barth to help Christians withstand the challenges of the Nazi party and of the "German Christians," a popular movement that saw no conflict between Christianity and the ideals of Hitler's National Socialism. It was thus a call to resistance against the theological claims of the Nazi state. The *Kairos* Document refers to the theological declaration made by a group of anonymous South African theologians based mostly in the townships of Soweto challenging the churches' response to the wicked policies of the Apartheid regime, especially as enunciated in the State of Emergency declared on July 21, 1985. Both documents could serve as models for the Church in Nigeria to emulate on the issue of oil-related injustice in the country.

### *8.2.5 Spirituality as a Form of Political Praxis: Prayer, Praise, and the Spoken Word*

One of the major misrepresentations of African evangelical, or 'Born Again', Christianity is the claim that it is a-political and uninterested with events and processes in the wider society. This has been shown not to be the case in this book. A major factor responsible for this misreading is the imposition of a western oriented dichotomy of 'the political' and 'the religious' on African Christianity. This imposition fails to take into account the way this form of Christianity perceives what constitutes the 'political' and its place and role in it. Such a reading views the task of theology regarding 'politics' from either the perspective of separation between theology and politics, on the one hand, or the perspective that sees theology as serving the cause of justice in the political realm, on the other (See Scott and Cavanaugh 2004, 2, for an analysis of the three predominant perspectives to the task of political theology; cf. Kirwan 2008, 5-9). Although there is a trace of these two views in the case studies, neither of them captures the predominant attitude of the three churches towards the political. Thus Ruth Marshall argues,

> Born-Again political rationalities, and the terms in which power, redemption, sovereignty, and other political themes are staged in its practices and professions of faith, stubbornly resist the distinction between sacred and secular we have come to take for granted in Western society and the categories of analysis we deploy to understand them (Marshall 2009, 3).

From the perspective of the three churches, the separation between the Church and the society is artificial since God as the *Ogologo Mma agha* is the absolute Sovereign of all of creation, including the so-called religious and secular orders. As *Tamuno*, *Papa*, *Baba*, Man-of-War, and Father of life, he is God-in-relationship who is just and good, and who enforces his justice by liberating and delivering his people from danger, by fighting for them, and by punishing the wicked, including oppressors and the perpetrators of injustice. Thus in the overall scheme of things, nothing or nobody is outside the reach of God.

Moreover, Jesus is regarded as Bulldozer and the ultimate *Oga-Kpatakpata* of all principalities and powers, all of which are accountable to him. Such powers include spiritual forces and human persons in places of authority. From their viewpoint, the work of the Holy Spirit is not restricted to indwelling and enduing believers with charismatic gifts but includes the cultivation of Christian character, the restoration of peace in a broken society, the enlightenment of the human mind, and the breaking of the yoke of militancy and cultism. Within this context, the church serves as the arena of expression of a different kind of politics, namely, spirituality expressed as political praxis and which is embedded in their worship, prayer, praise and the spoken word.

The seven roles of prayer in the spirituality of the three congregations were discussed in Chapter Five. They include the theological – prayer as expression of faith in God; the spiritual – prayer as a channel of penetrating and overcoming the diffused layers of spiritual forces; the political – prayer as a means of connecting to the realm of powers and enforcement of God's will through the spoken word, either as 'prophecy' or 'declaration'; and the physical – prayer as a sort of bullet proof. Others are the emotional – prayer as a means of succour for grief and loss in the face of conflict; the communal – prayer as a means of communal solidarity and support; and the social – prayer as a channel of intercession for peace. When these churches pray, they believe that they are not only performing a 'religious' function but also inherently unlocking the gates of evil forces militating against peace and progress in the society and thereby making an important political contribution since they believe that without their prayers the situation would have been worse. This belief is not exclusive to them as this writer found out. It was also expressed in separate interviews by Godwin Tasie, a retired Professor of Church History and at the time the Chairman of the Governing Council of The Rivers State University of Science and Technology, Port Harcourt; Apostle G.D. Numbere, the Chairman of Christian Association of Nigeria, South-South Zone; and Ledum Mitee, the President of MOSOP.

Not just their prayers, but also their entire liturgy is oriented toward political praxis. For example, Julian Ward has described some contributions of the Pentecostal movement, which are applicable to the churches in this study. They include the reality of the Pentecostal experience, namely, the baptism in the Holy Spirit; its lively worship, which engenders the prioritisation of spiritual needs and the promotion of godly living; its strong conviction that the gifts of the Holy Spirit are the norm for the Christian life; its belief that divine healing is a biblical reality; and its ability to attract, retain and mobilise the working class (Ward 1984, 203-204). The egalitarian ethos is promoted as members use their spiritual gifts and, by so doing, partake in the exercise of power and thereby acquire leadership skills. The acquisition of leadership skills is further strengthened through the various organisations and small groups in the churches where lay members, including women and youths, function as leaders. This ethos is a social leveller as it gives voice to those who otherwise would have been voiceless – women, particularly widows, youths, the poor, and the uneducated. When they worship, they create new networks of relationship (cf. Martin 1996, 44) and receive encouragement through praise, music, prayer, testimonies, and the spoken word of the preacher, which then serve as sources of inspiration and empowerment for confronting the vicissitudes of daily life. An example of this was given in Chapter Four in which a young architect in Faith Baptist shared the account of his deliverance from the bullet of youth gangs which he attributed to a declaration made by the Church Associate Pastor at a prayer meeting. Thus their spirituality is a legitimate form of political praxis. Once more, Marshall is right when she writes as follows:

> (T)he language of faith is truly performative and accords to testimony and prayer a force that is genuinely foundational: to invoke God, to praise Him, to pray to Him, and to testify to His works constitutes both the act of faith...but also, very literally, an *action* on the world. Through prayer and witness, converts do things with words (Marshall 2009, 4).

## 8.3 Crude Oil as Gift

This subsection develops from the fact that the predominant view in the case studies is that God is the Creator, Owner, Ruler, King and the One who blessed the Niger Delta region with crude oil, which they believe is intended for the good of all. Thus, the discussion that follows will formulate a theology/ethic of crude oil for Nigeria using Walter Brueggemann's biblical land ethic.

### *8.3.1 A Theological Perspective to Land: God as Owner and Giver of Oil*

There are a number of major economic views of natural resources. Two of them were highlighted in chapter three, namely, the neo-classical perspective, which sees mineral deposits essentially as capital assets, and the corporate viewpoint, which sees mineral deposits as sources of income. There is a third point of view, namely, the national or political perspective, which views minerals as national wealth and focuses its energy primarily on "the 'macro' context of such deposits" (Humphreys 1983, 171-172). A common denominator to all the three perspectives is the fact that they develop from a philosophical attitude that sees land as "absolute possession and property" (Bueggemann 2006, 30). This attitude is what lubricates the tendency towards acquisitiveness, which manifests in what Brueggemann calls eminent domain, confiscation, usurpation, and arrogant autonomy (Brueggemann 2006, 31-33; cf. the four 'foes' of the earth in the environmental crisis – human ignorance, human greed, human aggression and human arrogance – identified in Russell 1994, 72-93). Ecologically speaking, they are anthropocentric, utilitarian and instrumentalist in orientation regarding the natural order. This could be seen from the discussion on the key players in Nigeria's political economy of oil and conflict carried out in chapter three.

In contrast to these, Brueggemann offers a theological perspective to land, or "the land ethic in Scripture," in which land is seen as the creation of God for the good of all. Hence, he says, "The land, its potential for power, and its resources are to be devoted to the *common good*, that *all the neighbours* are to enjoy the fruitfulness and well-being of land as God's creation" (Brueggemann 2002, 3; cf. Von Rad 1966, 79-93). The discussion that follows will summarise and adapt Brueggemann's biblical land ethic in which he argues that land "is a central, if not *the central theme* of biblical faith" (Brueggemann 2002, 3), to crude oil in the Niger Delta.

### *8.3.2 A Biblical Land Ethic*

According to Brueggemann, a biblical land ethic has the following four affirmations: land as gift, land as temptation, land as task, and land as threat. He explains land as gift as follows: as gift from Yahweh, the Land of Promise binds Israel in new ways to him, the giver, since it did not acquire it "either by power or stratagem" (Brueggemann 2002, 47). Rather it was a gift based on Yahweh's radical grace. According to Brueggemann, the same land that is a source of life "has within it seductive power."

Hence, as temptation it "provides secured people with dangerous alternatives": either to maintain through memory an understanding of the source of the gift and by keeping the gift as gift, or by imagining "guaranteed security" on the land and seeing itself as the controller and manager of achievement rather than as creature of divine grace (Brueggemann 2002, 50). Land as task underscores the fact that the privilege of land gift has responsibility expressed in three ways: forbidding of images, keeping of the Sabbath, and care for the neighbour (Brueggemann 2002, 61-63). But land was also a threat to Israel on account of the more impressive Canaanites (Brueggemann 2002, 63). Israel is to enter into it in trust and bold confidence on the basis of Yahweh's promise; yet, in doing so, it is "to reject both easy accommodation and self-assured belligerence" (Brueggemann 2002 65).

Continuing, Brueggemann applies the theme of land to the NT in which he identifies a series of symbols that convey the motif, even though he acknowledges that geographical questions were not of primary concern to the writers of the NT (cf. Davies, 1994, 370). In the Synoptics, the symbol is the "Kingdom of God"; in John's Gospel, it is "eternal life"; in Paul, it is "inheritance"; while in Hebrews, it is "homeland." He then adds that the core symbol that cuts across all the traditions is the *crucifixion/resurrection*, which he states is dialectical and illuminated when linked to land (cf. Ezek 37:1-14). The dialectic is that of possessed land lost/exiles en route to the land of promise. Thus, crucifixion is a call to leave the old land (Mark 10:17-22), forgo power, and embrace the risk of powerlessness and landlessness. Put differently, it may be seen as land-loss so that others may receive the same land as gift. On the other hand, resurrection is the gift of power to the powerless and "the invitation to the dispossessed to enter new power, freedom, and life, that is, 'turf'." According to him, Jesus fully embodies this (Brueggemann 2002, 169-170). The next section will apply the above analysis of the biblical land ethic to the Niger Delta case by delineating an ethic of crude oil.

### *8.3.3 An Ethic of Crude Oil*

Just as land is described by Brueggemann as gift, temptation, task, and threat, crude oil in Nigeria could be described as gift, temptation, task, and threat.

### *Crude Oil as Gift*

From the perspective of the case studies, crude oil that is found in the Niger Delta is a divine endowment based on his providential care. Its presence in the region was only *discovered* in 1956 after decades of concerted search by foreign oil companies led by Shell-BP. Thus, the availability of crude oil in the Niger Delta is not based on human ingenuity, creativity, planning or merit. Rather, from the perspective of the case studies, it is a gift from God. What requires human effort and skill is its exploration, production and utilisation, whether at the 'upstream' or 'downstream' sectors. There are "four separate but vertically integrated functions" of the oil industry between the delivery of crude oil from the ground to the end-user, namely, production, transportation, refining, and marketing. The first function is the 'upstream' while the rest are the 'downstream' (Pearson 1970, 6).

### *Crude Oil as Temptation*

Just as in the case of the Israelites in which they yielded to the temptation of guaranteed satiation and covetousness in relation to land, the Nigerian nation lost the memory of its economic history and independence and fell prey to the pull of natural resource dependency and crude oil was turned into a national economic idol. Subsequently, two major abandonments occurred: first, with increasing oil revenue, especially following the first wave of oil boom in the 1970s, the agricultural sector that used to be the nation's main revenue base and foreign exchange earner was abandoned; and, second, the revenue formula that was hitherto used in which the regions received a greater share of the revenue from their regions was also discarded. For example, the Regional or State share of revenue according to the Revenue Allocation Formula enacted by the Federal Government of Nigeria was 100% in 1953, 50% in 1960, 45% in 1970, 20% in 1975, 25% in 1982, 1.55% in 1984; 3% in 1992; and 13% since 2000.

### *Crude Oil as Task*

If land is task to Israel, crude oil by analogy is also task to Nigeria. The three tasks – forbidding of images, keeping of the Sabbath, and care for the brother and sister – are also applicable to Nigerian political economy of oil. The tasks, which encompass the theological, the environmental, and the social, are critical in maintaining the integrity of the land and the continued existence of the nation as one people. Nigeria may lack a

Mosaic law like the Torah; yet it has its own traditions in the form of ethical principles, customs and rules of conduct preserved and passed on from one generation to another. These traditions explain the rationale, motivations, values and purpose of behaviour and subsequently serve as the "Scriptures," the people's source of knowledge about God's requirements (Kunhiyop 2008, 9), or "what the people must do to live ethically" (Idowu 1975, 42). In relation to crude oil, as the country yielded to the temptation of land, it reversed the three tasks above. Thus, in place of prohibition of idolatry, crude oil became a national idol; instead of land Sabbath, there is rapacious exploitation of oil resources leading to the degradation of land; and instead of care of the neighbour, there is greed and corruption.

### *Crude Oil as Threat*

In Nigeria's political economy, the threat of oil is that of arrogance and mismanagement of resources as a consequence of overdependence on it. One of the signs of the mismanagement is the pursuit of 'white elephant' projects by succeeding governments that encourage waste and corruption (Brooks 2004, 131). Also, the government is unwilling and/or unable to allocate funds for development. In addition it de-emphasises taxation and relies heavily on importation of goods to the detriment of local industries. Thus, despites its huge human and natural resources, Nigeria is rated by the World Bank as one of the poorest nations in the world per capita (World Bank 2006, 20). This is the threat of Nigeria's crude oil.

In summary, the above ethic of crude oil states that crude oil, which ultimately belongs to God, is a gift meant for the benefit of all, but Nigeria yielded to the temptation it posed, failed to effectively carry out its associated tasks, and therefore fell prey to its threat. This is the theological explanation of the resource curse. The panacea to this situation is a reversal of its present greed-infested, corruption-ridden, and elite-serving model of resource exploitation and management to a judicious, people-oriented one that transforms a resource curse into a resource blessing. The transformation has the capacity to eventually wean the country out of her over-dependence on the hydrocarbon industry.

Two major factors (economic and environmental) make this weaning necessary. The economic basis is the fact that oil accounts for a mere 11% of GNP and generates only 0.4% employment and benefits just 5% of the population (Brooks 2004, 128). The environmental basis of the need is captured by Nnimmo Bassey as follows: "Continued degradation in the form of spills and gas flares render the Niger Delta extremely vulnerable to the impacts of climate change with a projected loss of 50% ability to produce cereals by the year 2020 and would have suffered a 80% loss by 2050" (Bassey 2009, 1).

## 8.4 Total Freedom and Liberation Theology

The remaining challenge in this discussion is to relate it to liberation theology. The intention is not to give a detailed account and/or critique of the theology but to compare and contrast it with the above account of contextual political theology for the Niger Delta. Liberation theology, which originated in Latin America in the 1960s, is a broad and variegated cultural and ecclesial phenomenon in the Third World in which faith is confronted with oppression and injustice. Its beginning is linked to Vatican Council II (1961-1965) and its aftermath. According to John Milbank, Vatican Council II accorded social and political change more meaning than official Catholic dogma gave them, and had also placed emphasis on the social dimension of salvation (Milbank 2005, 913). It is also linked to Vatican Council II's emphases on the church as the people of God, the study of Scripture, and social justice (Berryman 1987, 12; Kee 1990, 131).

Liberation theology has a number of manifestations in Latin America, Africa, and Asia. Some of the major exponents of liberation theology in Latin America include Gustavo Gutiérrez, Juan Luis Segundo, Jon Sobrino, and Leonardo Boff. In Southern Africa where it is in the form of Black Theology, the leading voices include Desmond Tutu, Allan Aubrey Boesak, and Manas Buthelezi. The most widely known voice in the North American variant of Black Theology is James H. Cone. Some manifestations of the Asian form of liberation theology include *Minjung* Theology in Korea (See Bock 1981) and "the theology of struggle" in the Philippines (Wielenga 2007, 72). Among Feminist Theologians, four stand out, namely, R. Schussler Fiorenza, Rosemary Radford Ruether, Marcella Althaus-Reid, and Mercy Amba Oduyoye.

However, in whatever form it takes, liberation theology is a praxis-oriented protest against domination and against the socio-political systems and/or structures that facilitate them (cf. Metz and Schlick 1978, 48-49). For instance, in Latin America it is a protest against oppression and injustice by the ruling powers; in South Africa it served as a protest against white oppression and exploitation encoded in Apartheid; in America it is a protest against white racism; while feminist theologians protest against male domination.

There are three major strengths and contributions of liberation theology to the overall theological enterprise that are germane to this discussion. They are its emphasis and insistence that theology must be rooted in praxis, its preferential identification with the poor and oppressed, which Dyrness considers to be its most far-reaching and controversial contribution (Dyrness 1990, 90; cf. Gutiérrez 1983, ix; and Boff 1992, 3), and its introduction and use of the social sciences in its analysis. Thus, from the perspective of liberation theology, the gospel, sin and salvation are defined primarily in socio-political terms. For example, according to Gutiérrez, the political dimension of the gospel is seen in the fact that it is a liberator of the poor. He delineates three kinds of poverty, which include material poverty, which he terms scandalous; spiritual poverty, which he sees as openness to God and spiritual childhood; and poverty as commitment of solidarity with the poor and protest against oppression. He defines sin as the loss of the basic relationship of solidarity of God and human beings and consequently sees salvation more from a 'qualitative' perspective as "the communion of men with God and the communion of men among themselves…something which embraces all human reality, transforms it, and leads it to its fullness in Christ" (Gutiérrez 1983, 151-152). From this 'qualitative' perspective, therefore, salvation is not personal or otherworldly but rather collective and this-worldly. From his viewpoint the 'personal' and 'other-worldly' constitute the *quantitative* and *extensive* aspect of salvation, while restoration of communion between God and humans and among humans is the *qualitative* and *intensive* aspects (Gutiérrez 1983, 150-151). Furthermore, he postulates that the sphere in which this salvation is realized and justice restored is the Kingdom of God. In order to identify with the oppressed and give them unconditional love, Gutiérrez argues that one has to experience a "conversion" to them, or one's neighbour, which he sees as the source of Christian joy (Gutiérrez 1983, 194-203).

A question that arises from the foregoing is how does the theology of total freedom and social transformation relate with liberation theology in its variegated forms? As the following summary will show, they share commonalities as well as differences. First, the similarities: the two theologies share a number of common themes, such as focus on praxis, emphasis on love of neighbour, focus on social transformation founded on justice, and protest against injustice and oppression. Doctrinally they share common traits on a number of themes: (1) their respective Christologies with emphasis on historical and ecclesial Christology 'from below' (e.g., Sobrino 1978), in which Jesus is seen not only as Saviour but also as Liberator (Boff 1972, 43-46); (2) their soteriology, in which the social dimension of sin and the historical facet of salvation are taken seriously (e.g., see Gibellini 1986, 29); (3) their ecclesiology, in which the church is taken seriously as "the sign and the instrument of... salvation" (Gibellini) and as a centre of Christian praxis where everyone counts and is counted; and (4) their communal spirituality, which is seen as a form of commitment to Christ and a means of empowerment for social action (Gutiérrez 1984, 18-20). On account of the above similarities, it is appropriate, therefore, to state that the theology of total freedom and social transformation is in actual fact a form of liberation theology but which is presently a theology of *insistence* operating at the grassroots level, rather than a full-blown theology of *resistance*. Having said this, it is important to highlight a number of differences between the two forms of theology.

First, there are three major factors that are unique to the Niger Delta setting. They are its multi-religious make up, its multi-ethnic diversity, and the abundance of natural resources, particularly crude oil. Second, despite the fact that both forms of theology are connected to praxis, they differ in the order of movement between praxis and reflection. In liberation theology the order is praxis, reflection, and then renewal of praxis (Gutiérrez 1983, 6-15); but in the theology of total freedom, praxis, reflection and renewal move concurrently and influence each other. Third, unlike in liberation theology, the Exodus narrative does not feature in the theology of total freedom. Fourth, there is no appropriation of Marxist social hermeneutic. Rather the matrix of social composition in the Niger Delta is more complex than could be put into simple bi-polar categories. Finally, the two theologies differ on their respective emphasis on salvation. Whereas liberation theology stresses the 'qualitative' and this-worldly dimension, the theology of total freedom balances this with the 'quantitative' and otherworldly perspective.

## 8.5 Recommendations for Further Research

There are a number of issues that require further research in the light of this study. However, it will suffice to indicate some challenges and implications it has for both scholars and students of African Christianity, on the one hand, and the three congregations, on the other. First, for researchers on African Christianity, three prominent implications stand out: (1) In order to avoid a skewed appraisal and/or conclusions on the faith in its variegated expressions within the continent, scholars and students of African Christianity generally and African Christian theology in particular should continually grapple with the oral theology and daily praxis of African Christians – what they are saying and doing at the grassroots level in their everyday living and activities: through worship, singing, prayers, preaching, and teaching – and not depend solely on what their leaders and/or formal theologians are writing, preaching, teaching, and/or saying, as is often the case (cf. Longman 2010, 19; Hastings 1975, 347-361; Haynes 1996; Hyden 1986; Joseph 1993; and Pobee 1992).

For, as this study has shown, such an approach would help give a more complete picture of the way things are. This is why Timothy Longman aptly warns that a "mere recitation of the sermons, official letters, and other statements of church leaders ultimately says very little about the engagement of churches in society" (Longman 2010, 19; cf. Martin 2002, 89). (2) Several aspects of African evangelical Christianity have progressed beyond their missionary heritage, such as liturgy and theological expression, and therefore old labels and stereotypes ought to be re-evaluated in the light of the complex new realities as African Christians live out their faith and seek to theologically interpret it within their socio-political milieu. Hence the following warning by Christian Baeta in 1968 is most cogent: "We cannot artificially create 'African Theology' or even plan it: it must evolve spontaneously as the Church teaches and lives their faith in the extremely complex situation in Africa" (Baeta 1968, 332). And (3) there is the need to moderate the functionalist assessment of African evangelical Christianity as is presently the case with a view to taking more seriously the religious faith of its adherents and the way they view the world based on that faith, which Ruth Marshall has exemplified in her *Political Spiritualities*. It is not helpful to sustain the super-imposed, Euro-centric reading and classification of African Christianity.

The second set of challenges that arises from this study is for the three congregations and includes the following: (1) They need to improve and diversify their approach to Bible interpretation instead of solely depending on the literalist approach that was noted in the course of the investigation. Despite its pragmatic and existentialist advantages, it has the potential of misrepresenting the meaning and intention of particular passages as a result of failure to grapple with their contexts. J. Kwabena Asamoah-Gyadu points to this challenge when he describes it as a hermeneutical method that takes texts of scripture out of their context in order to support a preconceived notion (Asamoah-Gyadu 2005, 215). (2) The churches face the challenge of reorienting their membership on Christian loyalty and commitment so as to reduce the consumerist attitude towards church affiliation. And (3) Whereas Hilltop particularly should proactively take steps to preserve its current egalitarian ethos so as to avoid imbibing the authoritarian 'big-man' approach to church polity that has become associated with a number of African Pentecostal churches (see Gifford

2004, 175), the two Baptist churches, while maintaining their democratic credentials, should be careful not to sacrifice the need for community at the expense of individual expression, and vice versa.

On the issues that require further research, the following could be highlighted: First, there is the need for further empirical study of a similar nature in other churches – mainline, Pentecostal, AIC – so as to compare and contrast the findings of this research and further enrich the study. Second, further work is needed on a contextual political theology for the Niger Delta suggested in this book. A proposal to this effect has already been made by this author in "Redemptive Transformation: Towards a Theological Perspective to the Niger Delta Conflict," in the forthcoming book by Cambridge Scholars Publishing, *Moving Forward: Tradition and Transformation*. Third, and related to the above, there is need for further research into the capacity of Christian churches to serve as catalysts for enduring reconciliation, justice and peace in Nigeria's political economy, generally, and the oil industry, in particular, as this study could not dwell much on that due to reasons of space and focus.

# BIBLIOGRAPHY

## Select Books


Adogame, Afe and Lizo Jafta. "Zionists, Aladura and Roho: African Instituted Churches." In O.U. Kalu, J.W. Hofmeyr and P.J. Maritz, editors. *African Christianity: An African Story*. Pretoria: Department of Church History, University of Pretoria, 2005, 309-329.

Adogame, Afe, Roswith Gerloff and Klaus Hock, editors. *Christianity in Africa and the African Diaspora: The Appropriation of a Scattered Heritage*. London, New York: Continuum, 2008.

Agu, C.C. "The Primacy of Music in Igbo Traditional Religion." In Emma Ekpunobi and Ifeanyi Ezeaku, editors. *Socio-Philosophical Perspective of African Traditional Religion*, 49-55. Enugu: New Age Publishers, 1990.

Ajayi, J.F. Ade. *Christian Missions in Nigeria 1841-1891: The Making of a New Elite*. London: Longman, 1965.

Alagoa, Ebiegberi J. and Fombo Adadonye. *A Chronicle of Grand Bonny*. Ibadan: Ibadan University Press, 1972.

Alagoa, E.J. and Tekena N. Tamuno, editors. *Land and People of Nigeria: Rivers State*. Port Harcourt: Riverside Communications, 1989.

Alagoa, E.J. *A Short History of the Niger Delta: An Historical Interpretation of Ijo Oral Tradition*. Ibadan: Ibadan University Press, 1972.

Althaus-Reid, Marcella. *Liberation Theology and Sexuality*. London: Ashgate, 2006.

Ammerman, Nancy Tatom and others. *Congregation and Community*. New Brunswick, New Jersey: Rutgers University Press, 1997.

Anderson, Allan. *An Introduction to Pentecostalism: Global Charismatic Christianity*. Cambridge: Cambridge University Press, 2004.

Anderson, Martha G. "Bulletproof: Exploring the Warrior Ethos in Ijo Culture." In Martha G. Anderson and Philip M. Peek, editors. *Ways of the Rivers: Arts and Environment of the Niger Delta*. Los Angeles: UCLA Fowler Museum of Cultural History, 2002, 96-97.

Anderson, Martha G. and Philip M. Peek, editors. *Ways of the Rivers: Arts and Environment of the Niger Delta*. Los Angeles: UCLA Fowler Museum of Cultural History, 2002.

Ankrah, Kodwo E. "Church and Politics in Africa." In Kofi Appiah-Kubi and Sergio Torres, editors. *African Theology en Route*. Maryknoll: Orbis Books, 1977, 155-161.

Appiah-Kubi, Kofi and Sergio Torres, editors. *African Theology En Route*. Maryknoll, New York: Orbis Books, 1979.

Asamoah-Gyadu, J. Kwabena. *African Charismatics: Current Developments within Independent Indigenous Pentecostalism in Ghana*. Leiden: Brill, 2005.

—. "'I Will Put My Breath in You, and You Will Come to Life': Charismatic Renewal in Ghanaian Mainline Churches and Its Implications for African 'Diasporean' Christianity." In Afe Adogame, Roswith Gerloff and Klaus Hock, editors. *Christianity in Africa and the African Diaspora: The Appropriation of a Scattered Heritage*. London, New York: Continuum, 2008, 193-207.

Ayandele, E.A. *The Missionary Impact on Modern Nigeria 1842-1914: A Political and Social Analysis*. London: Longman, 1966.

Baeta, Christian. *Christianity in Tropical Africa: Studies Presented and Discussed at the Seventh International African Seminar*, University of Ghana, April 1965. London: Oxford University Press, 1968.

Barbé, Dominique. *A Theology of Conflict and Other Writings on Nonviolence*. Maryknoll: Orbis Books, 1977.

Barret, Lois. *The Way God Fights: War and Peace in the Old Testament*. Scottdale, Pa.: Herald Press, 1987.

Barth, Karl. *The Doctrine of the Word of God* (Church Dogmatics, I, 2nd edition). Trans. and eds. G.W. Bromiley and T.F. Torrance. Edinburgh: T & T Clark, 1956.

Batubo, A.B. "Baptist Church in Eastern Nigeria, 1893-1964." In J.A. Atanda, editor. *Baptist Churches in Nigeria 1850-1950: Accounts of Their Foundation and Growth*. Ibadan: University Press, 1988, 260-261.

Baur, John. *2000 Years of Christianity in Africa: An African History 62-1992*. Nairobi: Paulines Publications Africa, 1994.

Bayart, J .F. *The State in Africa: The Politics of the Belly*. London: Longmans, 1993.

Bediako, Kwame. *Christianity in Africa: The Renewal of a Non-Western Religion*. Edinburgh: Edinburgh University Press, 1997.

—. "Christian Witness in the Public Square: Some Lessons and Residual Challenges from the Recent Political History of Ghana." In Lamin Sanneh and Joel A. Carpenter, editors. *The Changing Face of Christianity*. New York: Oxford University Press, 2005, 117-132.

—. *Jesus and the Gospel in Africa: History and Experience*. Maryknoll:

Orbis Books, 2004.

Bennett, Zoe. "'Action is the Life of All': The Praxis-Based Epistemology of Liberation Theology." In Christopher Rowland, editor. *The Cambridge Companion to Liberation Theology*, 2nd ed. Cambridge: Cambridge University Press, 2007, 39-54.

Berryman, Phillip. *Liberation Theology: Essential Facts about the Revolutionary Movement in Latin America – and Beyond.* New York: Pantheon Books, 1987.

Best, S. and V. Kemedi. "Armed Groups and Conflict in Rivers and Plateau States Nigeria." In N. Florquin and E. Berman, editors. *Armed and Aimless.* Geneva: Small Arms Survey, 2005, 12-44.

Bevans, Stephen B. *Models of Contextual Theology.* Maryknoll: Orbis Books, 2000.

Bienen, Henry. "Nigeria: From: Windfall Gains to Welfare Losses." In Alan Gelb, editor. *Oil Windfalls: Blessing or Curse?* Oxford: Oxford University Press, 1988.

Bock, Kim Yong, editor. *Minjung Theology: People as the Subjects of History.* Singapore: Christian Conference of Asia, 1981.

Boesak, Allan Aubrey. *Black Theology, Black Power.* London and Oxford: Mowbrays, 1978.

—. *Farewell to Innocence: A Socio-Ethical Study on Black Theology and Black Power.* Maryknoll, New York: Orbis, 1977.

Boff, Leonardo. *Church: Charism and Power: Liberation Theology and the Institutional Church.* New York: Crossroad, 1985.

Boff, Leonardo, and Clodovis Boff. *Introducing Liberation Theology.* Translated by Paul Burns. Tunbridge Wells, Kent: Burns and Oates, 1992.

Boff, Leornado. *Jesus Christ Liberator.* Maryknoll, New York: Orbis Books, 1972.

Bonhoeffer, Dietrich. *Discipleship.* Edited by Martin Kuske and Ilse Todt. Minneapolis: Fortress Press, 2003.

—. *Ethics.* Edited by Clifford J. Green and translated by Reinhard Krauss, Charles C. West, and Douglas W. Stott. Minneapolis: Fortress Press, 2005.

Browning, Don S. *A Fundamental Practical Theology: Descriptive and Strategic Proposals.* Minneapolis: Fortress Press, 1991.

Browning, Don S., editor. *Practical Theology.* San Francisco: Harper and Row, 1983.

Brueggemann, Walter. *The Land: Place as Gift, Promise and Challenge in Biblical Faith.* Minneapolis: Augsburg Fortress Press, 2002.

—. *Theology of the Old Testament: Testimony, Dispute, Advocacy.*

Minneapolis: Augsburg Fortress Press, 1997.

Bujo, Benezet. *African Christian Morality at the Age of Inculturation.* Nairobi: Paulines Publications Africa, 1998.

Burgess, Richard. *Nigeria's Christian Revolution: The Civil War and its Pentecostal Progeny, 1967-2006*. Carlisle: Paternoster, 2008.

Buthelezi, Manas. "An African Theology or a Black Theology?" In Basil Moore, editor. *Black Theology: The South African Voice*. London: C. Hurst and Company, 1973, 29-35.

Cavanaugh, William T. *Torture and Eucharist: Theology, Politics, and the Body of Christ.* Oxford and Malden, Massachusetts: Blackwell Publishers, 1998.

Collins, Travis M. and S. Ademola Ishola. *Baptists and the Charismatic Movement.* Ibadan: Publishing Board, Nigerian Baptist Convention, 1995.

Cone, James H. *Black Theology and Black Power*. Maryknoll, New York: Orbis Books, 1969.

—. *A Black Theology of Liberation*, revised edition. Maryknoll, New York: Orbis Books, 1986.

—. *God and the Oppressed*. New York: The Seabury Press, 1975.

Cox, James L. and Gerrie ter Haar, editors. *Uniquely African? African Christian Identity from Cultural and Historical Perspectives*. Trenton, NJ: Africa World Press, 2003.

Cross, Anthony R. and Philip E. Thompson, editors. *Baptist Sacramentalism.* Carlisle: Paternoster Press, 2003.

Csordas, Thomas J. *Body/Meaning/Healing*. New York: Palgrave Macmillan, 2002.

—. "Embodiment and Cultural Phenomenology." In Gail Weiss and Honi Fern Haber, editors. *Perspectives on Embodiment: The Intersections of Nature and Culture*. London: Routledge, 1999.

—. *Language, Charisma, and Creativity: The Ritual of Life of a Religious Movement*. Berkeley and Los Angeles; London: University of California Press, 1997.

Dappa-Biriye, Harold J. R. *Minority Politics in Pre- and Post-Independence Nigeria.* Choba, Port Harcourt: University of Port Harcourt Press, 1995.

Davies, W.D. *The Gospel and the Land: Early Christianity and Jewish Territorial Doctrine.* Sheffield: Sheffield Academic Press, 1994.

Davis, Stephen T., Daniel Kendall and Gerald O'Collins, editors. *The Redemption: An Interdisciplinary Symposium on Christ as Redeemer.* Oxford: Oxford University Press, 2004.

De Gruchy, John W. and Charles Villa-Vicencio, editors. *Doing Theology in Context: South African Perspectives*. Maryknoll, NY: Orbis Books, 1994.

De Gruchy, John W. *Christianity and Democracy: A Theology for a Just World Order*. Cambridge: Cambridge University Press, 1995.

—. *Reconciliation: Restoring Justice*. London: SCM Press, 2002.

Dickson, Kwesi A. and Paul Ellingworth, editors. *Biblical Revelation and African Beliefs*. London: Lutterworth Press, 1969.

Dickson, Kwesi A. *Theology in Africa*. London: Darton, Longman and Todd; Maryknoll, New York: Orbis Books, 1984.

Dike, K. Onwuka. *Trade and Politics in the Niger Delta 1830-1885: An Introduction to the Economic and Political History of Nigeria*. Oxford: Clarendon Press, 1959.

Dyrness, William A., editor. *Emerging Voices in Global Christian Theology*. Grand Rapids: Zondervan, 1994.

—. *Learning about Theology from the Third World*. Grand Rapids: Academie Books, 1990.

Ela, Jean-Marc. *African Cry*. Maryknoll: Orbis Books, 1986.

—. *My Faith as an African*. Maryknoll: Orbis Books, 1988.

Environmental Rights Action/Friends of the Earth Nigeria. *Gas Flaring in Nigeria: A Human Rights, Environmental and Economic Monstrosity*. Amsterdam: Environmental Rights Action/Friends of the Earth Nigeria, 2005.

Fabella, Virginia and Mercy Oduyoye, editors. *With Passion and Compassion: Third World Women Doing Theology*. Maryknoll: Orbis Books, 1988.

Falola, Toyin and Julius O. Ihonvbere, editors. *Nigeria and the International Capitalist System*. Boulder and London: Lynne Rienner Publishers, 1988.

Federal Republic of Nigeria. *Niger Delta Regional Development Master Plan*.

Fleshman, M. "The International Community and the Crises in the Oil Producing Communities." In W. Raji, A. Ale, and E. Akinsola, editors., *Boiling Point* (Lagos: CDHR, 2000), 180-190.

Fiorenza, R. Schussler. *Discipleship of Equals*. New York: Crosswords, 1993.

Fisher, Simon and others. *Working with Conflict: Skills and Strategies for Action*. London: Zed Books, 2000.

Forrest, Tom. *Politics and Economic Development in Nigeria*. Boulder, Colorado: Westview Press, 1993.

Frankfort-Nachmias, Chava and David Nachmias. *Research Methods in the Social Sciences*, 5th Edition. London: Arnold, 1997.

Freeman, Curtis W., James Wm. McClendon, Jr., and C. Rosalee Velloso da Silva, editors. *Baptist Origins: A Reader in the Theology of a Christian People*. Valley Forge, PA: Judson Press, 1999.

Frynas, Jedrzej Georg. *Oil in Nigeria: Conflict and Litigation between Oil Companies and Village Communities*. London: LIT VERLAG, 1999.

Gehman, R.J. *Doing African Christian Theology: An Evangelical Perspective*. Nairobi: Evangel Publishing House, 1987.

Gelb, Alan H. *Oil Windfalls: Blessing or Curse?* Oxford: Oxford University Press, 1988.

Gibellini, Rosino. *The Liberation Theology Debate*. London: SCM Press, 1986.

—. Editor. *Paths of African Theology*. London: SCM Press, 1994.

Gifford, Paul. *African Christianity: Its Public Role*. London: Hurst and Company, 1998.

—. Editor. *Christian Churches and the Democratisation of Africa*. Leiden, New York, and Koln: E.J. Brill, 1995.

—. *Christianity and Politics in Doe's Liberia*. Cambridge: Cambridge University Press, 1993.

—. *Christianity: To Save or Enslave?* Harare: Ecumenical Documentation and Information Centre of Eastern and Southern Africa (EDICESA), 1990.

—. *Ghana's New Christianity: Pentecostalism in a Globalising African Economy*. London: Hurst & Company, 2004.

—. "A View of Ghana's New Christianity." In Lamin Sanneh and Joel A. Carpenter, editors. *The Changing Face of Christianity*. New York: Oxford University Press, 2005, 81-96.

Glaser, Barney G. and Anselm L. Strauss. *The Discovery of Grounded Theory: Strategies for Qualitative Research*. New York: Aldine Publishing, 1980.

Gottwald, Norman. *The Tribes of Yahweh*. Maryknoll, N.Y.: Orbis, 1979.

Gutierrez, Gustavo. *A Theology of Liberation: History, Politics and Salvation*. Translated and edited by Caridad Inda and John Eagleson. London: SCM, 1983.

—. *We Drink from Our Own Wells*. Maryknoll, New York: Orbis Books, 1984.

Habel, Norman C. *The Land is Mine: Six Biblical Land Ideologies*. Minneapolis: Fortress Press, 1995.

Hammett, John S. *Biblical Foundations for Baptist Churches: A Contemporary Ecclesiology*. Grand Rapids: Kregel Publications, 2005.

Harvey, Barry. "Re-Membering the Body: Baptism, Eucharist and the Politics of Disestablishment." In Anthony R. Cross and Philip E. Thompson, editors. *Baptist Sacramentalism – Studies in Baptist History and Thought, Vol. 5*. Carlisle: Paternoster Press, 2003, 96-97.

Hastings, Adrian. *African Christianity: An Essay in Interpretation*. London: Geoffrey Chapman, 1976.

—. *A History of African Christianity, 1950-1975*. Cambridge: Cambridge University Press, 1979.

Hauerwas, Stanley. *A Community of Character: Toward a Constructive Christian Social Ethic*. Notre Dame: University of Notre Dame Press, 1986.

—. *The Peaceable Kingdom: A Primer in Christian Ethics*. Notre Dame: University of Notre Dame Press, 1986.

Hauerwas, Stanley and William H. Willimon. *Resident Aliens: A Life in the Christian Colony*. Nashville: Abingdon Press, 1992.

Haynes, Jeff. *Religion and Politics in Africa*. Nairobi: East African Educational Publishers; London and New Jersey: Zed Books, 1996.

Heitink, Gerben. *Practical Theology: History, Theory, Action Domains: Manual for Practical Theology*. Translated by Render Bruinsma. Grand Rapids: Wlliam B. Eerdmans Publishing Company, 1999.

Helmick, Raymond G. and Rodney L. Petersen, editors. *Forgiveness and Reconciliation: Religion, Public Policy and Conflict Transformation*. Philadelphia and London: Templeton Foundation Press, 2001.

Human Rights Watch. *Nigeria: Crackdown in the Niger Delta*. New York: Human Rights Watch, 1999.

—. *Chop Fine: The Human Rights Impact of Local Government Corruption and Mismanagement in Rivers State, Nigeria*, vol. 19, no. 2(A), January 2007.

Hyden, G. "Governance and the Study of Politics in Africa." In G. Hyden & M. Bratton, editors, *Governance and Politics in Africa*. Boulder, CO: Lynne Rienne, 1992.

Idowu, E. Bolaji. *African Traditional Religion: A Definition*. Maryknoll: Orbis Books, 1975.

—. *Towards an Indigenous Church*. London: Oxford University Press, 1965.

Ihonvbere, Julius O. and Timothy M. Shaw. *Towards a Political Economy of Nigeria: Petroleum and Politics at the (Semi-) Periphery*. Aldershot: Avebury, 1988.

Ikein, Augustine A. and Comfort Briggs-Anigboh. *Oil and Fiscal Federalism in Nigeria: The Political Economy of Resource Allocation in a Developing Country*. Aldershot: Ashgate, 1998.

Ikein, Augustine A. *The Impact of Oil on a Developing Economy: The Case of Nigeria*. New York: Praeger Publishers, 1990.

Ikenga-Metuh, Emefie. *African Religions in Western Conceptual Schemes: The Problem of Interpretation.* Ibadan: Pastoral Institute, 1985.

—. *God and Man in African Religion: A Case Study of the Igbo of Nigeria*, 2nd ed. Enugu: SNAAP Press, 1999.

Ikime, Obaro. *Niger Delta Rivalry: Itsekiri-Urhobo Relations and the European Presence 1884-1936*. London and Harlow: Longmans, 1969.

Isichei, Elizabeth *A History of Christianity in Africa: From Antiquity to the Present*. London: SPCK, 1995.

James, Wilmot and Linda van de Vyer, editors. *After the TRC: Reflections on Truth and Reconciliation in South Africa*. Athens: Ohio University Press, 2001.

Jenkins, Philip. *The New Faces of Christianity: Believing the Bible in the Global South*. Oxford: Oxford University Press, 2006.

Jones, G.I. *From Slaves to Palm Oil: Slave Trade and Palm Oil Trade in the Bight of Biafra*. Cambridge: African Studies Centre, 1989.

—. *The Trading States of the Oil Rivers: A Study of Political Development in Eastern Nigeria*. London, Ibadan, Accra: Oxford University Press, 1963.

Joseph, Richard. "The Christian Churches and Democracy in Contemporary Africa." In John Witte, editor, *Christianity and Democracy in Global Context*. Boulder: Westview, 1993, 231-247.

Kalu, O.U., editor. *African Christianity: An African Story*. Pretoria: University of Pretoria, Department of Church History, 2005.

—. *African Pentecostalism: An Introduction.* Oxford: Oxford University Press, 2008.

—. *Clio in a Sacred Garb: Essays on Christian Presence and African Responses, 1900-2000*. Trenton, NJ: Africa World Press, 2008.

—. *The History of Christianity in West Africa*. London: Longman, 1980.

Katongole, Emmanuel and Chris Rice. *Reconciling All Things: A Christian Vision for Justice, Peace and Healing*. Downers Grove: IVP Books, 2008.

Katongole, Emmanuel, editor. *African Theology Today*. Scranton: The University of Scranton Press, 2002.

Kee, Alistair. *Marx and the Failure of Liberation Theology*. London: SCM Press; Philadelphia: Trinity Press International, 1990.

Khan, Sarah Ahmad. *Nigeria: The Political Economy of Oil*. Oxford: Oxford University Press, 1994.

Kirwan, Michael. *Political Theology: A New Introduction*. London: Darton, Longman and Todd, 2008.

Krueger, Richard A. *Focus Group: A Practical Guide for Applied Research*, 2nd Edition. London: SAGE Publications, 1994.

Kukah, Matthew Hassan. "Christians and Nigeria's Aborted Transition." In Paul Gifford, editor, *The Christian Churches and the Democratisation of Africa*. Leiden: E.J. Brill, 1995, 228-229.

—. *Democracy and Civil Society in Nigeria.* Ibadan: Spectrum Books, 2000.

Kunhiyop, Samuel Waje. *African Christian Ethics*. Nairobi: Word Alive Publishers; Bukuru: African Christian Textbooks; Grand Rapids: Zondervan, 2008.

Land, Steven J. *Pentecostal Spirituality: A Passion for the Kingdom.* Sheffield: Sheffield Academic Press, 1994.

Lederach, John Paul. *Building Peace: Sustainable Reconciliation in Divided Societies.* Washington, D.C.: United States Institute of Peace Press, 2004.

—. *The Journey Toward Reconciliation.* Scottdale: Herald Press, 1999.

—. *The Little Book of Conflict Transformation.* Intercourse, PA: Good Books, 2003.

—. *Preparing for Peace: Conflict Transformation across Cultures.* New York: Syracuse University Press, 1995.

Lind, Millard C. *Yahweh is a Warrior*. Scotdale, Pa.: Herald, 1980.

Litvin, Daniel. *Empires of Profit: Commerce, Conquest and Corporate Responsibility.* New York: Texere, 2003.

Locke, Hubert G, editor. *The Barmen Confession: Papers from Seattle Assembly*. Lewiston, NY: The Edwin Mellen Press, 1986.

Longman, Timothy. *Christianity and Genocide in Rwanda*. Cambridge: Cambridge University Press, 2010.

Magesa, Laurenti. *African Religion: The Moral Traditions of Abundant Life*. Maryknoll: Orbis Books, 1997.

Mana, Kä. *Christians and Churches of Africa: Salvation in Christ and Building of a New African Society*. Maryknoll: Orbis Books, 2004.

Manby, Bronwen. *The Price of Oil: Corporate Responsibility and Human Rights Violations in Nigeria's Oil Producing Communities*. New York: Human Rights Watch, 1999.

Marshall, Ruth. "'God is not a Democrat': Pentecostalism and Democratisation in Nigeria." In Paul Gifford, editor, *The Christian Churches and the Democratisation of Africa*. Leiden: E.J. Brill, 1995.

—. *Political Spiritualities: The Pentecostal Revolution in Nigeria.* Chicago: The University of Chicago Press, 2009.

Martey, Emmanuel. *African Theology: Inculturation and Liberation.* Maryknoll: Orbis Books, 1993.

Martin, David and Peter Mullen, editors. *Strange Gifts? A Guide to Charismatic Renewal*. Oxford: Basil Blackwell Publisher, 1984.

Martin, David. *Forbidden Revolutions: Pentecostalism in Latin America and Catholicism in Eastern Europe*. London: SPCK, 1996.

—. *Pentecostalism: The World Their Parish.* Oxford: Blackwell Publishers, 2002.

—. *Reflections on Sociology and Theology*. New York: Oxford University Press, 1997.

Maxwell, David. "African Gifts of the Spirit: Fundamentalism and the Rise of the Born-Again Movement in Africa." In Martyn Percy and Ian Jones, editors, *Fundamentalism, Church and Society*. London: SPCK, 2002, 160-182.

May, Rollo. *Power and Innocence: A Search for the Sources of Violence*. New York: W.W. Norton and Company, 1972.

Mbachu, Dulue. *The Poverty of Oil Wealth in Nigeria's Delta*. Zurich: International Relations and Security Network, 2006.

Mbiti, John S. *African Religions & Philosophy*. London, Ibadan, Nairobi: Heinemann, 1969.

—. *Bible and Theology in African Christianity.* Nairobi: OUP, 1986.

—. "The Biblical Basis for Present Trends in African Theology." In K. Appiah-Kubi and S. Torres, editors, *African Theology En Route*. Maryknoll, New York: Orbis Books, 1979, 83-94.

McClendon, James Wm Jr. *Doctrine*. Nashville: Abingdon Press, 1994.

—. *Ethics*. Nashville: Abingdon Press, 1986.

—. *Witness*. Nashville: Abingdon Press; Bristol: Alban, 2000.

McGrath, Alister E. *Christian Theology: An Introduction*. Oxford: Blackwell Publishers, 1997.

McKeown, Robert L. *In the Land of the Oil Rivers: The Story of the Qua Iboe Mission*. London: Marshall Brothers, 1902.

Metz, Rene and Jean Schlick, editors. *Liberation Theology and the Message of Salvation: Papers of the Fourth Cerdic Colloquim Strasbourg, May 10-12, 1973*. Translated by David G. Gelzer. Pittsburgh, Pennsylvania: The Pickwick Press, 1978.

Moltmann, Jürgen. *The Spirit of Life: A Universal Affirmation*. Translated by Margaret Kohl. Minneapolis: Fortress Press, 1992.

—. *The Theology of Hope*. New York: Harper and Row, 1967.

Mugambi, J.N.K. *African Christian Theology: An Introduction*. Nairobi: Heinemann Kenya, 1989; reprint 1990.

—. *From Liberation to Reconstruction: African Christian Theology after the Cold War*. Nairobi: East African Educational Publishers, 1995.

Muzorewa, G.H. *The Origins and Development of African Theology.* Maryknoll: Orbis Books, 1985.

Na'Allah, Abdul-Rasheed, editor. *Ogoni's Agonies: Ken Saro Wiwa and the Crisis in Nigeria.* Trenton, NJ: Africa World Press, 1998.

Nadeau, Kathleen M. *Liberation Theology in the Philippines: Faith in a Revolution.* Westport, CT: Praeger Publishers, 2002.

National Population Commission (NPC), Nigeria and ORC Macro. *Nigeria Demographic and Health Survey 2003.* Calverton, Maryland: National Population Commission and ORC Macro, 2004.

Nnadozie, Emmanuel U. *Oil and Socioeconomic Crisis in Nigeria: A Regional Perspective to the Nigerian Disease and the Rural Sector.* Lewiston: Mellen University Press, 2005.

Nodu, Asoa K. *Baptist Work in Ogoni.* Port Harcourt: Awesome Resources Nig., 2004.

Northcott, Michael S. *An Angel Directs the Storm: Apocalyptic Religion and American Empire.* London: I.B. Tauris, 2004.

—. *A Moral Climate: The Ethics of Global Warming.* London: Darton, Longman and Todd, 2007.

—. *The Environment and Christian Ethics.* Cambridge: Cambridge University Press, 2001.

Numbere, Nonyem E. *Greater Evangelism World Crusade: Landmarks in Her Vision and History.* New and Enlarged edition. Port Harcourt: Greater Evangelism World Crusade Publications, 2002.

Nyamiti, Charles. *Christ as Our Ancestor: Christology from an African Perspective.* Gweru, Zimbabwe: 1984.

—. "Contemporary African Christologies: Assessment and Practical Suggestions." In Rosino Gibellini, editor, *Paths of African Theology.* Maryknoll: Orbis Books, 1994.

—. *The Scope of African Theology.* Kampala: Gaba Publication, 1973.

Oduro, Thomas. "Church Music in the Life of African Christian Communities." In Roberta King and others, editors, *Music in the Life of the African Church.* Waco: Baylor University Press, 2008.

Oduyoye, Mercy Amba. *Beads and Strands: Reflections of an African Woman on Christianity in Africa.* Carlisle: Regnum and Cle in association with Paternoster Press; Yaounde: Regnum, Cle, 2002.

—. *Hearing and Knowing: Theological Reflections on Christianity in Africa.* Maryknoll: Orbis Books, 1986.

Ogan, Charles. *Unity of a People: Search for Peace in Okrika, Rivers State, Nigeria.* Port Harcourt: Charles Ogan, 1988.

Okonta, Ike and Oronto Douglas. *Where Vultures Feast: Shell, Human Rights, and Oil in the Niger Delta.* San Francisco: Sierra Club Books,

2001.

Onoh, J.K. *The Nigerian Oil Economy: From Prosperity to Glut*. Kent: Croom Helm, 1983.

OPEC. *Annual Statistical Bulletin 2004*. Vienna: OPEC, 2005.

—. *Monthly Oil Market Report January 2006*. Vienna: OPEC, 2006.

O'Donovan, Oliver. *The Desire of the Nations: Rediscovering the Roots of Political Theology*. Cambridge: Cambridge University Press, 2003.

—. *Resurrection and Moral Order: An Outline for Evangelical Ethics*. Second Edition. Leicester: Apollos, 1994.

Olayiwola, Peter O. *Petroleum and Structural Change in a Developing Country: The Case of Nigeria*. New York: Praeger Publishers, 1987.

Orobator, Agbonkhianmeghe E. *Theology Brewed in an African Pot*. Maryknoll: Orbis Books, 2008.

Osuigwe, Nkem. "Paul Gifford on African Evangelical and Fundamentalist Christianity: An Ethnographic Evaluation." In Heather M. Morgan, Jernej Letnar Cernic and Lindsay Milligan, editors, *Perspectives on Power: An Inter-disciplinary Perspective*. Newcastle: Cambridge Scholars Publishing, 2010, 288-302.

—. "Redemptive Transformation: Towards a Theological Perspective to the Niger Delta Conflict." In *Moving Forward: Tradition and Transformation*. London: Anthem Press, 2011, forthcoming.

Osuoka, Asume. *The Shell Report: Continuing Abuses in Nigeria 10 Years after Ken Saro-Wiwa*. Benin City: Environmental Rights Action (ERA)/Friends of the Earth, 2005.

Parratt, John, editor. *An Introduction to Third World Theologies*. Cambridge: Cambridge University Press, 2005.

—. *A Reader in African Christian Theology*. London: SPCK, 1990.

—. *Reinventing Christianity: African Theology Today*. Grand Rapids: William B. Eerdmans Publishing Company and Trenton, New Jersey: Africa World Press, 1995.

Pearson, Scott R. *Petroleum and the Nigerian Economy*. Stanford, California: Stanford University Press, 1970.

Perham, Margery. *Lugard: The Years of Authority, 1898-1945*. London: Collins, 1961.

Pobee, J. *Religion and Politics in Ghana: A Case Study in the Acheampong Era, 1972-1979*. Accra: Ghana Universities Press, 1992.

Powell, Mark A. *Jesus as a Figure in History*. Louisville, KY: Westminster John Knox Press, 1998.

Redekop, Calvin, editor. *Creation and the Environment: An Anabaptist Perspective on a Sustainable World*. Baltimore, Maryland: The Johns Hopkins University Press, 2000.

Reid, Harry. *Reformation: The Dangerous Birth of the Modern World*. Edinburgh: Saint Andrew Press, 2009.

Richardson, Alan. *The Political Christ*. London: SCM Press, 1973.

Royal Dutch/Shell Group of Companies. *Financial and Organizational Information 2000-2004*. London: Royal Dutch/Shell Group of Companies, 2005.

Ruether, Rosemary Radford. *Sexism and God-Talk: Toward a Feminist Theology*. London: SCM, 1983.

Russell, Colin A. *The Earth, Humanity and God*. London: UCL Press, 1994.

Sanneh, Lamin and Joel A. Carpenter, editors. *The Changing Face of Christianity: Africa, the West, and the World*. Oxford: Oxford University Press, 2005.

Sanneh, Lamin. *West African Christianity: The Religious Impact*. London: C. Hurst & Company, 1983.

Santmire, H. Paul. *The Travail of Nature: The Ambiguous Ecological Promise of Christian Theology*. Minneapolis: Fortress Press, 1985.

Saro-Wiwa, Ken. *Nigeria: The Brink of Disaster*. Port Harcourt: Saros International Publishers, 1991.

Sawyerr, Harry. *Christology in Creative Evangelism: Towards a New Christian Encounter with Africa*. London: Lutterworth Press, 1968.

Schatzl, L.H. *Petroleum in Nigeria*. London: Oxford University Press, 1969.

Scobie, Charles H.H. *The Ways of Our God: An Approach to Biblical Theology*. Grand Rapids; Cambridge: William B. Eerdmans Publishing Company, 2003.

Scott, Peter and William T. Cavanaugh, editors. *The Blackwell Companion to Political Theology*. Oxford: Blackwell, 2004.

Segundo, Juan Luis. *The Community Called Church*. Translated by John Drury. Dublin: Gill and Macmillan, 1980.

—. *Liberation of Theology*. Translated by John Drury. Dublin: Gill and Macmillan, 1977.

Shelley, Bruce and Marshall Shelley. *Consumer Church: Can Evangelicals Win the World without Losing Their Souls?* Downers Grove, Ill.: InterVarsity, 1992.

Shorter, A. *African Christian Theology*. Maryknoll: Orbis Books, 1986.

Sindima, Harvey J. *Drums of Redemption: An Introduction to African Christianity*. Westport, CT: Praeger, 1994.

Slatter, Michael. "The Schleitheim Confession." In Curtis W. Freeman, James Wm. McClendon, Jr., and C. Rosalee Velloso da Silva, editors,

*Baptist Origins: A Reader in the Theology of a Christian People.* Valley Forge, PA: Judson Press, 1999.

Sobrino, Jon. *Christology at the Crossroads: A Latin American Approach.* Maryknoll, New York: Orbis, 1978.

Stinton, Diane. "Africa, East and West." In John Parratt, editor, *An Introduction to Third World Theologies.* Cambridge: Cambridge University Press, 2005, 105-136.

Stinton, Diane B. *Jesus of Africa: Voices of Contemporary African Christology.* Maryknoll: Orbis Books, 2006.

Stückelberger, Christoph and J.N.K. Mugambi, editors. *Responsible Leadership: Global and Contextual Ethical Perspectives.* Geneva: WCC Publications, 2007.

Sundkler, Bengt and Christopher Steed. *A History of the Church in Africa.* Cambridge: Cambridge University Press, 2001.

Ter Haar, Gerrie and James J. Busuttil, editors. *Bridge or Barrier: Religion, Violence and Visions for Peace.* Leiden: Brill, 2005.

Tierney, Thomas F. "The Preservation and Ownership of the Body." In Gail Weiss and Honi Fern Haber, editors, *Perspectives on Embodiment: The Intersections of Nature and Culture.* London: Routledge, 1999.

Torrance, Alan J. and Michael Banner, editors. *The Doctrine of God and Theological Ethics.* London and New York: T & T Clark International, 2006.

Truett, George Washington. "Baptists and Religious Liberty, 1920." In Joseph Early, Jr., *Readings in Baptist History: Four Centuries of Selected Documents.* Nashville: B & H Publishing Group, 2008.

Tutu, Desmond Mpilo. *Hope and Suffering: Sermons and Speeches.* London: Fount Paperbacks, 1986.

—. *No Future without Forgiveness.* London: Rider, 2000.

Uche, Chibuike U. and Ogbonnaya C. Uche. *Oil and the Politics of Revenue Allocation in Nigeria.* Leiden: African Studies Centre, 2004.

Villa-Vicencio, Charles. *A Theology of Reconstruction: Nation-Building and Human Rights.* Cambridge: Cambridge University Press, 1992.

—. *Trapped in Apartheid: A Socio-Theological History of the English-Speaking Churches.* Maryknoll: Orbis Books, 1988.

Volf, Miroslav. *Exclusion and Embrace: A Theological Exploration of Identity, Otherness, and Reconciliation.* Nashville: Abingdon Press, 1996.

Von Rad, Gerhard. *Holy War in Ancient Israel.* Translated by Marva J. Dawn. Grand Rapids: Eerdmans, 1991.

—. *The Problem of the Hexateuch and Other Essays*. Translated by E.W. Trueman Dicken. New York: McGraw-Hill, 1966.

Walls, A. F. "Africa and the Future of Christianity: A Summary and Reflection." In Christopher Fyfe and Andrew F. Walls, editors, *Christianity in Africa in the 1990s*. Edinburgh: Centre for African Studies, University of Edinburgh, 1996, 139-148.

—. "Introduction: African Christianity in the History of Religions." In Christopher Fyfe and Andrew Walls, editors, *Christianity in Africa in the 1990s*. Edinburgh: Centre of African Studies, University of Edinburgh, 1996, 1-16.

Ward, Julian. "Pentecostal Theology and the Charismatic Movement." In David Martin and Peter Mullen, editors, *Strange Gifts? A Guide to Charismatic Renewal*. Oxford: Basil Blackwell, 1984, 192-216.

West, Gerald O. and Musa W. Dube, editors. *The Bible in Africa: Transactions, Trajectories, and Trends*. Leiden: Brill, 2000.

Westermann, Claus. *What Does the Old Testament Say About God?* Atlanta: John Knox Press, 1979.

White, R.E.O. *Biblical Ethics*. Leominster: Gracewing, 1994.

Wielenga, Bastiaan. "Liberation Theology in Asia." In Christopher Rowland, editor, *The Cambridge Companion to Liberation Theology*, 2nd ed. Cambridge: Cambridge University Press, 2007, 55-78.

Wink, Walter. *Engaging the Powers: Discernment and Resistance in a World of Domination*. Minneapolis: Fortress Press, 1992.

—. *Jesus and Nonviolence: A Third Way*. Minneapolis: Fortress Press, 2003.

Wiwa, Ken. *In the Shadow of a Saint*. London: Doubleday, 2000.

Wogaman, J. Philip. *Christian Perspectives on Politics*. London: SCM Press, 1998.

World Bank. *Sub-Saharan Africa: From Crisis to Sustainable Growth.* Washington, DC: World Bank, 1989.

—. *Where is the Wealth of Nations?* Washington, DC: The International Bank for Reconstruction and Development/The World Bank, 2006.

World Council of Churches (WCC). *For a New Africa with Hope and Dignity*. Geneva: WCC Publications, 2004.

—. *The Nature and Mission of the Church: A Stage on the Way to a Common Statement*. Geneva: WCC, 2005.

Yoder, John Howard. *Body Politics: Five Practices of the Christian Community before the Watching World*. Scottdale: Herald Press, 2001.

—. *The Christian Witness to the State*. Scottdale: Herald Press, 2002.

—. *The Original Revolution: Essays on Christian Pacifism*. Scottdale: Herald Press, 1972.

—. *The Politics of Jesus: Vicit Agnus Noster*. 2nd ed. Grand Rapids; Carlisle: Eerdmans and Paternoster Press, 1994.

—. *The Priestly Kingdom: Social Ethics as Gospel*. Notre Dame: University of Notre Dame Press, 1984.

—. *The Schleitheim Confession*. Scottdale, Pennsylvania; Waterloo, Ontario: Herald Press, 1977.

Young III, Josiah U. *African Theology: A Critical Analysis and Annotated Bibliography*. Westport: Greenwood Press, 1993.

## Select Dictionary and Encyclopedia Articles

Akinade, Akintunde E. "African Initiated Churches." *Encyclopedia of Pentecostal and Charismatic Christianity*, 2006 edition.

Bundy, D.D. "European Pietist Roots of Pentecostalism." *The New International Dictionary of Pentecostal and Charismatic Movement*, revised and expanded, 2002 edition.

Burgess, Stanley. "Neocharismatic Movements." *Encyclopedia of Pentecostal and Charismatic Christianity*, 2006 edition.

Doak, Brian R. and William P. Griffin. "Anointing with Oil." *Encyclopedia of Pentecostal and Charismatic Christianity*, 2006 edition.

Fiorenza, Francis Schüssler. "Redemption." *The New Dictionary of Theology,* 1987 edition.

Jeong, Ho-Won. "Conflict Management and Resolution." *Encyclopedia of Violence, Peace, and Conflict*, Vol. 1, 1999 edition.

Kevan, E.F. "The Baptist Tradition." *Evangelical Dictionary of Theology*, 1984 edition.

Kriesberg, Louis. "Conflict Transformation." *Encyclopedia of Violence, Peace, and Conflict*, vol. 1, 1999.

Lechner, Frank J. "Fundamentalism." *Encyclopedia of Religion and Society*, 1998 edition.

Milbank, John. "Liberation Theology." *Encyclopedia of Christian Theology*, Vol. 2, 2005 edition.

Ojo, Matthews A. "African Charismatics." *Encyclopedia of African and African-American Religions*, 2001 edition.

Okoye, James C. "African Theology." *Dictionary of Mission: Theology, History, Perspectives*, 1997 edition.

Okure, Teresa. "Bible: Africa." *Dictionary of Third World Theologies*, 2000 edition.

Olupona, J.K. "Global Survey: West Africa." *The New International Dictionary of Pentecostal and Charismatic Movements*, 2002 Revised and Expanded Edition.

Parker, J.I. "Authority." *New Bible Dictionary*, 2nd edition, 1988.

## Select Interviews

Dimorgu, Gunn, Pioneer Pastor of Faith Baptist Church, Port Harcourt. Interview by author, Ahoada, October 26, 2006.

George, C.T.T. Interview by author, Port Harcourt, November 9, 2006.

Konya, Roseline, Rivers State Commissioner for Environment. Interview by author, Port Harcourt, September 5, 2006.

Mitee, Ledum, President, MOSOP. Interview by author, Port Harcourt, August 17, 2006.

Numbere, Apostle G.D., Chairman, South-South Zone, Christian Association of Nigeria. Interview by author, Port Harcourt, September 21, 2006.

Princewill, T.J.T., HRM, King Amachree XI, *Amanyanabo* of Kalabari Kingdom. Interview by author, Port Harcourt, August 30, 2006.

Tasie, G.O.M. Chairman, Governing Council, Rivers State University of Science and Technology. Interview by author, Port Harcourt, August 29, 2006.

## Select Journal Articles

Adams, Nicholas and Charles Elliot. "Ethnography is Dogmatics: Making Description Central to Systematic Theology." *Scottish Journal of Theology*, 53 (Autumn, 2000), 339-364.

Adom-Oware, Seth. "'Who Do You Say that I Am?' (Matt. 16:15): A Contemporary African Response." *West African Journal of Ecclesial Studies* (WAJES), No. 3 (December, 1991), 9-23.

Ajayi, Kunle. "From the Demand for Sovereign National Conference to National Dialogue: The Dilemma of the Nigerian State." *Stud. Tribes Tribals*, Vol. 4, No. 2 (2006), 126.

Akinade, Akintunde E. "'Who Do You Say that I Am?' – An Assessment of Some Christological Constructs in Africa." *Asia Journal of Theology*, Vol. 9, No. 1 (April, 1995), 181-200.

Aniagwu, John K.A. "The Theology and Practice of Reconciliation Revisited." *West African Journal of Ecclesial Studies* (WAJES), No. 3 (December, 1991), 37-55.

Appiah-Kubi, Kofi. "Indigenous African Christian Churches: Signs of Authenticity." *Bulletin of African Theology,* Vol. 1, No. 2 (July-Dec., 1979), 241-249.

Asamoah-Gyadu, K. "Of Faith and Visual Alertness: The Message of 'Mediatized' Religion in an African Pentecostal Context." *Material Religion*, Vol. 1, No.3 (2005), 336–357.

Balcomb, Anthony O. "Left, Right and Centre: Evangelicals and the Struggle for Liberation in South Africa." *Journal of Theology of Southern Africa*, 118 (March, 2004), 146-160.

Bayinsana, Eugene. "Christ as Reconciler in Pauline Theology and in Contemporary Rwanda." *Africa Journal of Evangelical Theology*, Vol. 15, No. 1 (1996), 19-28.

Bediako, Kwame. "Africa and Christianity in the Threshold of the Third Millennium: The Religious Dimension." *African Affairs,* 99 (2000), 303-323.

—. "Jesus in African Culture." *Evangelical Review of Theology*, Vol. 17, No. 1 (January, 1993), 54-64.

—. "The Doctrine of Christ and the Significance of Vernacular Terminology." *International Bulletin of Missionary Research*, Vol. 22, No. 3 (July, 1998), 110-111.

—. Book Review. *The Christian Churches and the Democratisation of Africa*, by Paul Gifford. *African Affairs*, Vol. 96, No. 384. (July, 1997), 468-469.

Bourne, Richard. "Witness, Democracy and Civil Society: Reflections on John Howard Yoder's Exilic Ecclesiology." *Ecclesiology,* 3 (2007), 195-213.

Brueggemann, Walter. "To Whom Does the Land Belong? 2 Samuel 3:12." *Journal for Preachers*, Vol. 30, No. 3 (2006), 30.

Brunnschweiler, Christa N. "Cursing the Blessings? Natural Resource Abundance, Institutions, and Economic Growth." *World Development*, Vol. 36, No. 3 (2008), 399–419.

Burgess, Richard. "Freedom from the Past and Faith for the Future: Nigerian Pentecostal Theology in Global Perspective." *PentecoStudies*, vol. 7, No. 2 (2008), 29-63.

Burleson, Blake Wiley. "John Mbiti's Theology as a Reflection of the Archaic Notion of Corporate Personality." *Africa Theological Journal*, Vol. 21, No. 2 (1992), 164-187.

Cahalan, Kathleen A. "Three Approaches to Practical Theology, Theological Education, and the Church's Ministry." *International Journal of Practical Theology*, 9 (2005), 63-94.

Coast, Ernestina. "An Evaluation of Demographers' Use of Ethnographies." *Population Studies*, Vol. 57, No. 3 (November, 2003), 337-346.

Connolly, Hugh G. "Conversion: A Lifelong Journey." *The Furrow*, Vol. 44, No. 3 (March, 1993), 159-166.

Conradie, Ernst M. "Towards an Ecological Reformulation of the Christian Doctrine of Sin." *Journal of Theology for Southern Africa*, 122 (July, 2005), 4-22.

Croy, N. Clayton. "The Messianic Whippersnapper: Did Jesus Use a Whip on People in the Temple (John 2:15)?" *Journal of Biblical Literature*, Vol. 128, No. 3 (Fall, 2009), 555-568.

Daly, M. Catharine, Joanne B. Eicher and Tonye V. Erekosima. "Male and Female Artistry in Kalabari Dress." *African Arts*, Vol. 19, No. 3 (May, 1986), 48-51, 83.

De Carvalho, Emilio J. M. "What Do the Africans Say That Jesus Christ Is?" *Africa Theological Journal*, Vol. 10, No. 2 (1981), 17-25.

—. "Who is Jesus Christ for Africa Today?" *Africa Theological Journal*, Vol. 10, No. 1 (1981), 27-36.

De Witte, Marleen. "Accra's Sounds and Sacred Spaces." *International Journal of Urban and Regional Research*, Volume 32, No. 3 (September, 2008), 690–709.

—. "Altar Media's Living Word: Televised Charismatic Christianity in Ghana." *Journal of Religion in Africa*, Vol. 33, No. 2 (2003), 172–202.

Dwane, Sigqibo. "Christology in the Third World." *Journal of Theology for Southern Africa*, 21 (December, 1977), 3-12.

Eberlein, Ruben. "On the Road to the State's Perdition? Authority and Sovereignty in the Niger Delta, Nigeria." *Journal of Modern African Studies*, Vol. 44, No. 4 (2006), 573-596.

Ekeh, Peter. "Report on the Peace Summit on the Warri Crisis Washington, DC, July 24, 1999." *ACAS Bulletin*, No. 68 (Fall, 2004), 34.

Ellis, Stephen and Gerrie ter Haar. "Religion and Politics in Sub-Saharan Africa." *The Journal of Modern African Studies*, 36, 2 (1998), 176.

Emole, Chijioke E. "The Petroleum (Amendment) Decree, 1996, of Nigeria." *Journal of African Law*, Vol. 41, No. 2(1997), 239-245.

Enyinnaya, John. "The Church as a Clan: An Ecclesiology for the Igbo of Southeast Nigeria," *BETFA*, 1 (2002), 27-41.

—. "A Theological Response to Contemporary Worship Patterns in Nigerian Churches." *BETFA: Journal of the Ogbomoso Circle*, No. 2 (2003), 18-28.

Erekosima, Tonye Victor and Joanne Bubolz Eicher. "Kalabari Cut-Thread and Pulled-Thread Cloth." *African Arts*, Vol. 14, No. 2 (February, 1981), 48-51+87.

Friedman, Thomas L. "The First Law of Petropolitics." *Foreign Policy* (May/June, 2006), 28-36.

Gifford, Paul. "'Africa Shall Be Saved': An Appraisal of Reinhard Bonnke's Pan-African Crusade." *Journal of Religion in Africa*, Vol. 17, Fasc. 1 (February, 1987), 63-92.

—. "Christian Fundamentalism and Development." *Review of African Political Economy,* Vol. 19 No. 52, Fundamentalism in Africa: Religion and Politics (November, 1991), 9-20.

—. "Ghana's Charismatic Churches Source." *Journal of Religion in Africa*, Vol. 24, Fasc. 3 (August, 1994), 241-265.

—. "Persistence and Change in Contemporary African Religion." *Social Compass*, Vol. 51, No. 2 (2004), 169-176.

—. "Some Recent Developments in African Christianity." *African Affairs*, Vol. 93, No. 373 (October, 1994), 513-534.

Haar, Gerrie Ter. Book Review. *Ghana's New Christianity: Pentecostalism in a Globalising African Economy*, by Paul Gifford. *African Affairs*, vol. 105, No. 420 (July, 2006), 486-487.

Hackett, Rosalind I. J. "Charismatic/Pentecostal Appropriation of Media Technologies in Nigeria and Ghana." *Journal of Religion in Africa*, Vol. 28, Fasc. 3. (August, 1998), 258-277.

Hastings, Adrian. "Christianity and Revolution." *African Affairs*, 74 (July, 1975), 347-361.

Haynes, Jeff. "Religion and Democratization in Africa." *Democratization*, Vol.11, No.4 (August, 2004), 66–89.

Heimbrock, Hans-Gunter. "From Data to Theory: Elements of Methodology in Empirical Phenomenological Research in Practical Theology." *International Journal of Practical Theology*, 9 (2005), 273-299.

Horton, Robin. "The Kalabari 'Ekine' Society: A Borderland of Religion and Art." *Africa: Journal of the International African Institute*, Vol. 33, No. 2. (April, 1963), 94-114.

—. "The Kalabari World-View: An Outline and Interpretation." *Journal of the International African Institute*, vol. 32, No. 3 (July, 1962), 197-220.

Humphreys, David. "The Value of Mineral Resources: Perspectives and Conflicts." *Resources Policy* (September, 1983), 169-176.

Hyden, Goran. "Religion, Politics, and the Crisis in Africa." *UFSI Reports*, 1986, no. 18.

Ibeanu, Okechukwu. "Janus Unbound: Petro-business and Petropolitics in the Niger Delta." *Review of African Political Economy*, no. 91, vol. 29 (2002), 164-165.

—. "The Rhetoric of Rights: Understanding the Changing Discourses of Rights in the Niger Delta." *ACAS Bulletin*, No. 68 (Fall, 2004), 15-19.

Ibeanu, Okey. "(Sp)oils of Politics: Petroleum, Politics and the Illusion of Development in the Niger Delta, Nigeria." *ACAS Bulletin*, No. 64 (2002/2003), 16-36.

Ifeka, Caroline. "Youth Cultures & the Fetishization of Violence in Nigeria." *Review of African Political Economy*, Vol. 33, No. 110 (September, 2006), 721-736.

Ikelegbe, Augustine. "Civil Society, Oil and Conflict in the Niger Delta Region of Nigeria: Ramifications of Civil Society for a Regional Resource Struggle." *Journal of Modern African Studies*, vol. 39, no. 3 (2001), 437-469.

—. "The Perverse Manifestation of Civil Society: Evidence from Nigeria." *The Journal of Modern African Studies*, Vol. 39, No. 1 (2001), 1-24.

Ikenga-Metuh, Emefie. "Religious Concepts in West African Cosmogonies: A Problem of Interpretation." *Journal of Religion in Africa*, Vol. 13, Fasc. 1. (1982), 11-24.

Imasogie, Osadolor. "The Influence of African Traditional Religious Ideas of Worship on the Christian Worship Practices in Nigeria." *Ogbomoso Journal of Theology*, 6 (December, 1991), 17-23.

Ingersoll, Fern and Jasper Ingersoll. "Both a Borrower and a Lender Be: Ethnography, Oral History, and Grounded Theory." *The Oral History Review*, Vol. 15, No. 1 Fieldwork in Oral History (Spring, 1987), 81-102.

Janesick, Valerie J. "Constructing Grounded Theory: Reflections on a Case Study of a Professor of Architectural Design." *Studies in Art Education*, Vol. 24, No. 1 (1982), 16-24.

Jonker, W.D. "The Credibility of the Church." *Missionalia*, Vol. 7, No. 3 (November, 1979), 114-127.

Joseph, Richard A. "Affluence and Underdevelopment: The Nigerian Experience." *The Journal of Modern African Studies*, Vol. 16, No. 2 (June, 1978), 221-239.

Kanieski, Mary Ann. Book Review. *Congregation and Community*, by Nancy T. Ammerman. *Review of Religious Research*, Vol. 39, No. 2, Special Issue: Media and Unconditional Religion (December, 1997), 188.

Kasambala, Amon Eddie. "The Impact of an African Spirituality and Cosmology on God-Images in Africa: A Challenge to Practical

Theology and Pastoral Ministry." *International Journal of Practical Theology*, Vol. 9 (2005), 300-323.

Kemedi, Dimieari von. "The Changing Predatory Styles of International Oil Companies in Nigeria." *Review of African Political Economy*, Vol. 30, No. 95 (March, 2003), 134-135.

King, Hazel. "Cooperation in Contextualization: Two Visionaries of the African Church – Mojola Agbebi and William Hughes of the African Institute, Colwyn Bay." *Journal of Religion in Africa*, Vol. 16, No. 1 (February, 1986), 2-21.

Le Billon, Philippe. "The Political Ecology of War: Natural Resources and Armed Conflicts." *Political Geography*, Vol. 20 (2001), 561–584.

Ludwig, Frieder. "Elijah II: Radicalisation and Consolidation of the Garrick Braide Movement 1915-1918." *Journal of Religion in Africa*, Vol. 23, Fasc. 4. (November, 1993), 296-317.

Madujibeya, S.A. "Oil and Nigeria's Economic Development." *African Affairs*, Vol. 75, No. 300 (July, 1976), 284-316.

Maimela, Simon S. "Traditional African Anthropology and Christian Theology." *Journal of Theology for Southern Africa*, 76 (September, 1991), 4-14.

Makumbe, J. "Is there a Civil Society in Africa?" *International Affairs*, Vol. 74, No. 2 (1998), 305-317.

Manby, Bronwen. "Oil Jihad in the Niger Delta?" *Open Democracy*, 2, October 7, 2004.

Marshall, Ruth. "Power in the Name of Jesus." *Review of African Political Economy,* 52, (1991), 21-38.

Marshall-Fratani, Ruth. "Mediating the Global and Local in Nigerian Pentecostalism." *Journal of Religion in Africa*, Vol. 28, Fasc. 3 (August, 1998), 278-315.

Maxwell, David. "Delivered from the Spirit of Poverty?': Pentecostalism, Prosperity and Modernity in Zimbabwe." *Journal of Religion in Africa*, Vol. 28, Fasc. 3 (August, 1998), 350-373.

—. "In Defence of African Creativity." Book Review of *African Christianity: Its Public Role*, by Paul Gifford. *Journal of Religion in Africa*, Vol. 30, Fasc. 4 (November, 2000), 468-481.

Mayall, James. "Oil and Nigerian Foreign Policy." *African Affairs*, Vol. 75, No. 300 (July, 1976), 317-330.

Mbiti, John. "Cattle Are Born with Ears, Their Horns Grow Later." *African Theological Journal*, Vol. 8, No. 1 (1979), 15-25.

—. "Christianity and African Culture." *Evangelical Review of Theology*, Vol. 3, No. 2 (October, 1979), 183-197.

Meagher, Kate. "Hijacking Civil Society: The Inside Story of the Bakassi

Boys Vigilante Group of South-Eastern Nigeria." *Journal of Modern African Studies*, Vol. 45, No. 1 (2007), 89-115.

Meredith, David. "Government and the Decline of the Nigerian Oil-Palm Export Industry, 1919-1939." *The Journal of African History*, Vol. 25, No. 3 (1984), 311-329.

Meyer, Birgit. "The Power of Money: Politics, Occult Forces, and Pentecostalism in Ghana." *African Studies Review*, Vol. 41, No. 3 (December, 1998), 15-37.

Munungi, John. "The Academy and the Crisis of African Governance." *African Issues* (Fall, 2004), 9-23.

Musasiwa, Roy B. "The Finality of Jesus in Africa." *Evangelical Review of Theology*, Vol. 17, No.1 (January, 1993), 65-69.

Muzorewa, Gwinyai. "Christ as Our Ancestor: Christology from an African Perspective' by Charles Nyamiti: A Review Essay." *Africa Theological Journal*, Vol. 17, No. 3 (1988), 255-64.

Naanen, Ben. "The Political Economy of Oil and Violence in the Niger Delta." *ACAS Bulletin*, No. 68 (Fall, 2004), 4-9.

Nkwoka, A. O. "Jesus as Eldest Brother, (Okpara): An Igbo Paradigm for Christology in African Context." *Asia Journal of Theology*, Vol. 5, No. 1 (1991), 87-103.

Nzomiwu, John P.C. "Toward an Understanding of the Igbo Traditional Perception of Justice." *Journal of African Studies*, 15:1/2 (Spring/Summer, 1988), 33-39.

Odion-Akhaine, Sylvester. "Liberation Theology in Nigeria?" *Review of African Political Economy*, No. 110, Vol. 33 (September, 2006), 760.

Ojo, Matthews A. "The Contextual Significance of the Charismatic Movements in Independent Nigeria." *Africa: Journal of the International African Institute*, Vol. 58, No. 2. (1988), 175-192.

Okonmah, Patrick D. "Right to a Clean Environment: The Case for the People of Oil-Producing Communities in the Nigerian Delta." *Journal of African Law*, Vol. 41, No. 1 (1997), 43-67.

Okonta, Ike. "Death-Agony of a Malformed Political Order." *ACAS Bulletin*, No. 68 (Fall, 2004), 23-26.

Omeje, Kenneth. "Petrobusiness and Security Threats in the Niger Delta, Nigeria." *Current Sociology*, 54 (2006), 477-498.

Omoweh, Daniel A. "Is it the Warri Crisis or the Crisis of the Nigerian State?" *ACAS Bulletin*, No. 68 (Fall, 2004), 10-15.

Onyeidu, S.O. "Confession in the Traditional Religions of Africa." *West African Religion*, Vol. 16, No. 1 (1975), 8-14.

Orobator, Emmanuel. "Perspectives and Trends in Contemporary African Ecclesiology." *Studia Missionalia*, 45 (1996), 267-82.

Osuigwe, Nkem. "An African Christian Theology of Reconciliation." *BETFA: Journal of the Ogbomoso Circle*, 1 (2002), 1-15

Patterson, A.S. "A Reappraisal of Democracy in Civil Society: Evidence from Rural Senegal." *Journal of Modern African Studies*, Vol. 36, No. 2 (1998), 423-441.

Perami, Judith. "Pelete-Bite: Kalabari Cut-Thread Cloth." *African Arts*, Vol. 16, No. 1 (November, 1982), 80-81.

Peter, C.B. "African Hyphenated Christians an Alternate Model of Theologizing in Africa." *Nordic Journal of African Studies*, Vol. 3, No. 1 (1994), 100–113.

Pobee, John S. "Political Theology in the African Context." *Africa Theological Journal*, Vol. 11, No. 2 (1982), 68-75.

Sankey: Paul J. "The Church as Clan: Critical Reflections on African Ecclesiology." *International Review of Mission*, Vol. 83, No. 330 (July, 1994), 437-49.

Schatz, Sayre P. "Nigeria's Petro-Political Fluctuation." *Issue: A Journal of Opinion*, Vol. 11, no. 1/2 (Spring-Summer, 1981), 36.

Senior, Donald. "The New Testament and Peacemaking: Some Problem Passages." *Faith and Mission* (Fall, 1986), 76.

Smith, Daniel Jordan. "'The Arrow of God': Pentecostalism, Inequality, and the Supernatural in South-Eastern Nigeria." *Africa: Journal of the International African Institute*, Vol. 71, No. 4. (2001), 587-613.

Stevens, Paul and Evelyn Dietsche. "Resource Curse: An Analysis of Causes, Experiences and Possible Ways Forward." *Energy Policy*, 36 (2008), 56–65.

Talbot, P. Amaury. "Some Beliefs of Today and Yesterday (Niger Delta Tribes)." *Journal of the Royal African Society*, Vol. 15, No. 60 (July, 1916), 305-319.

Tasie, G.O.M. "Priesthood and Social Change in the Traditional Religion of the Kalabari People." *Journal of Religion in Africa*, Vol. 3, Fasc. 1 (1970), 96-105.

Thomas, Norman E. "Images of Church and Mission in African Independent Churches." *Missiology*, Vol. 23, No. 1 (January, 1995), 17-29.

Total E&P Nigeria. "Total's New Sustainable Development Vision." *Partnership*, No. 11 (January-March, 2009), 7.

Ukeje, C. "From Aba to Ugborodu." *Oxford Development Studies*, Vol. 32, No. 4 (2004), 605-617.

Ukpong, J. S. "Developments in Biblical interpretation in Africa: Historical and Hermeneutic Directions." *Journal of Theology in Southern Africa,* 108 (2000), 3-18.

—. "The Political Dimension of Jesus' Ministry – Implications for Evangelization in Africa." *West African Journal of Ecclesial Studies* (WAJES), Vol. 3, No. 1 (1995), 1-10.

Ward, Kevin. Book Review. *African Christianity: Its Public Role*, by Paul Gifford. *African Affairs*, Vol. 98, No. 391 (April, 1999), 272-274.

Wendland, Ernst. "'Who Do People Say that I Am?' Contextualizing Christology in Africa." *Africa Journal of Evangelical Theology*, Vol. 10, No. 2 (1991), 13-32.

Westley, William A. "The Escalation of Violence through Legitimation." *Annals of the American Academy of Political and Social Science*, Vol. 364: Patterns of Violence (March, 1966), 120-126.

Wilson, John. Book Review. *Congregation and Community*, by Nancy Tantom Ammerman. *Contemporary Sociology*, Vol. 26, No. 6 (November, 1997), 769-770.

Wiseman, John. "Aspects of Social Organisation in a Nigerian Petrol Queue." *The Journal of Modern African Studies*, Vol. 17, No. 2 (June, 1979), 317-323.

Zalik, Anna. "The Niger Delta: 'Petro-violence' and 'Partnership Development.'" *Review of African Political Economy*, 31:101 (2004), 404.

## Select Reports, Lectures and Other Documents

*2006-2007 Catholic Directory*. Iperu-Remo, Ogun State: AMBASSADOR Publications, 2006.

Abubakar, Mohammed and Emeka Anuforo. "NEITI Declares N768m Oil Revenue Missing." *The Guardian*, Lagos, http://www.guardiannewsngr.com/news/article04 (accessed August 14, 2007).

Amadi, Sam. "Theological Ambivalence and Democratic Accountability." Paper presented at "Pentecostal-Civil Society Dialogue" organized by CLASA at the Muson Centre on Monday, October 18, 2004.

Anderson, Allan. "Evangelism and the Growth of Pentecostalism in Africa": http://artsweb.bham.ac.uk/aanderson/Publications/evangelism_and_the_growth_of_pen.htm (accessed May 30, 2007).

—. "The Globalization of Pentecostalism." Paper presented at the 2002 Commission Meeting of the Churches' Commission on Mission, September 14-16, 2002, Bangor/Wales.

Bassey, Nnimmo, Executive Director, Environmental Rights Action (ERA)/Friends of the Earth Nigeria (FoEN) and Chair, FoE

International. "Oil, Environment and Human Rights – The Case of Nigeria's Niger Delta." Paper presented at the Quaker Meeting House, Edinburgh, November 21, 2009.

Bediako, Kwame. "Cry Jesus! Christian Theology and Presence in Modern Africa." The 1993 Laing Lecture at the London Bible College, Friday February 5, 1993.

Diamond, L. "Prospects for Democratic Development in Africa." Paper presented at the Democratic Governance Project, Department of Political and Administrative Studies, University of Zimbabwe, Harare, 1997.

Effiong, Joe. *The Sunnewsonline,* Lagos, "Nabbed! Gang Leader Whose Group Terrorises All," http://www.sunnewsonline.com/webpages/news/national/2006/oct/10/national-10-10-2006-09.htm (accessed October 10, 2006).

Ekeinde, Austin. "Interview: Most Nigerian Rebels Take Amnesty – Ex-Militant." Reuters, September 27, 2009, http://www.reuters.com/article/africaCrisis/idUSLR97607 (accessed November 16, 2009).

Emmanuel, O. Emmanuel, "Gauging Progress on Oil in Nigeria: Community Relations, Development Impact and Revenue Transparency." Paper presented at a Panel Discussion on "The Crude Reality: Africa's Oil Boom and the Poor" at Fordham University School of Law, New York, September 27, 2004.

Energy Information Administration (EIA), "Nigeria Country Analysis."

Eso, Justice Kayode, Chairman, Rivers State Truth and Reconciliation Commission. Speech on the Occasion of the Submission of the Report of the Truth and Reconciliation Commission on Tuesday 10th March 2009, *The Punch*, Tuesday, March 17, 2009.

Fabowale, Yinka and Isaac Anumihe. "How NNPC fleeced Nigeria of N555bn, by RMAFC," *Sunnewsonline*, Lagos, August 13, 2007, http://www.sunnewsonline.com/webpages/news/national/2007/aug/13/national-13-08-2007-001.htm (accessed August 13, 2007).

Federal Ministry of Finance, Federal Republic of Nigeria, Abuja. *Distribution of Revenue Allocation to State Governments by Federation Account Allocation Committee*. For various months.

Fischer, Martina, Berghof Research Center for Constructive Conflict Management. "Recovering from Violent Conflict: Regeneration and (Re-)Integration as Elements of Peacebuilding," May 2004.

Fleshman, Michael D. "Nigeria: Blood and Oil." A Special Report Prepared for the Africa Fund, New York, September 16, 1999.

Ginifer, J. and O. Ismail, "Armed Violence and Poverty in Nigeria." Centre for International Cooperation and Security, Working Paper, University of Bradford, UK, 2005.

Hilltop International Christian Centre, Port Harcourt. "Reaching the Nations with Christ's Love," n.d.

Human Rights Watch. "Politics as War: The Human Rights Impact and Causes of Post-Election Violence in Rivers State, Nigeria," Vol. 20, no 3(A) (March, 2008).

—. "Rivers and Blood: Guns, Oil and Power in Nigeria's Rivers State." Human Rights Watch Briefing Paper, February 2005.

—. "'They Do Not Own This Place': Government Discrimination against "Non-Indigenes" in Nigeria," Volume 18, No. 3(A) (April, 2006).

—. "Violence in Nigeria's Oil Rich Rivers State in 2004: A Human Rights Watch Briefing Paper," February 2005.

—. "The Warri Crisis: Fueling Violence," November 2003.

Hyden, Goran, the University of Florida, USA. "Why Africa Finds it so Hard to Develop," http://www.giga-hamburg.de/openaccess/nordsued aktuell/2004_4/giga_nsa_2004_4_hyden.pdf. (accessed November 27, 2009).

Ijaw Youth Council. "Kaiama Declaration."

Ikwunze, C., E. Amaize, and E. Ogoigbe. "Bayelsa: Return Home, Obi begs Fleeing Kaiama Residents." *Vanguard*, Lagos, January 6, 1999, 1-2.

Institute for Peace and Conflict Resolution, The Presidency, Federal Government of Nigeria. "Strategic Conflict Assessment: Nigeria." Consolidated Report, October 2002.

Itsueli, U.J. "The Petroleum Industry: Nigeria's Extracted Gift." Paper delivered at the NEITI Workshop in Abuja, February 15, 2005.

Iyayi, Festus. "Transparency and the Oil Dilemma." Paper Presented at the 2nd Anniversary Lecture of the Nigeria Extractive Industries Transparency Initiative, the Congress Hall, Transcorp Hilton Hotel, Abuja, Friday February 17, 2005.

Kehinde, Dipo. "Police Nab Port Harcourt Cultist." *The Sun News*, Lagos, online version, Wednesday, August 29, 2007.

Marshall, Katherine. "Africa: How and Why is Faith Important and Relevant for Development?" Draft unpublished paper, April 30, 2005.

Movement for the Survival of Ogoni People (MOSOP). *Ogoni Bill of Rights.*

Niger Delta Congress. "Declaration of Niger Delta Bill of Rights."

—. The Niger Delta Manifesto.

Nigeria Extractive Industry Transparency Initiative. *Final Audit: Interim Report on Financial Flows 2003-2004*. Presented by Hart Nurse Ltd and SS Afemikhe & Co. to the National Stakeholder Working Group on 8th January, 2006

Nigerian Baptist Convention. *Statements of Faith and Practice*. As Approved by the Nigerian Baptist Convention-in-Session at Igede-Ekiti on April 29, 1993.

Nwilo, P.C. and O.T. Badejo. "Impacts of Oil Spills along the Nigerian Coast," http://www.aehsmag.com/issues/2001/october/impacts.htm (accessed November 9, 2005).

Odion, Louis. "Mend vs. Igbinedion & Ibori," *The Sunnewsonline*, Lagos, Sunday, February 11, 2007, http://www.sunnewsonline.com/webpages/columnists/bottomline/bottomline-february-11-2007.htm (accessed February 11, 2007).

Oguine, Ike. "Nigeria's Oil Revenue and the Oil Producing Areas." http://www.dundee.ac.uk/cepmlp/journal/html/vol14/article4-10.html. (accessed February 23, 2006).

Ofiebor, Okafor. "Rival Cult Groups Vow to Spill Blood in Rivers State," http://www.legaloil.com/NewsItem.asp?DocumentIDX=1091238785&Category=news (accessed October 17, 2006).

Ogoigbe, Emmanuel. "Day of Terror: Youths Inflict Bodily Injuries on Family Members over Illegal Levy." *The Sunnewsonline*, Lagos, Tuesday January 30, 2007: http://www.sunnewsonline.com/webpages/news/national/2007/jan/30/national-30-01-2007-05.htm (accessed January 30, 2007).

Ojo, Mattews A. "Pentecostalism, Public Accountability and Governance in Nigeria." Paper presented at a Workshop on "Pentecostal-Civic Society Dialogue on Public Accountability and Governance" on Monday October 18, 2004 at the Agip Recital Hall, MUSON Centre, Onikan, Lagos, Nigeria.

Okafor, C. "Group Advises Oil Firms to withdraw Workers." *The Guardian*, November 23, 1999.

Omiyi, Basil. "Shell Nigeria Corporate Strategy for Ending Gas Flaring." Paper Presented at a Seminar on Gas Flaring and Poverty Alleviation in Oslo, Norway, June 18-19, 2001.

Onah, George. "Bodo, Mogho Bloodbath in Ogoni." V*anguard*, Lagos, http://www.vanguardngr.com/articles/2002/cover/march07/03032007/f303032007.html (accessed March 3, 2007).

Osaghae, Eghosa E. and Rotimi T. Suberu. "A History of Identities, Violence, and Stability in Nigeria." Centre for Research on Inequality,

Human Security and Ethnicity, CRISE, University of Oxford, Working Paper No. 6, January 2005, 5-13.

Osuigwe, Nkem. "Nigeria and Her Current Oil Windfall." Sermon Preached at Faith Baptist Church, Port Harcourt on Sunday August 29, 2004.

—. "The Political Economy of Oil Related Violence in Nigeria: The Birth of Quasi Sovereignties and its Challenges for the Church." Paper Presented at the Scottish Universities Conference for Postgraduate Students in Divinity and Religious Studies Held at King's College, University of Aberdeen, Thursday June 21, 2007.

—. "Quest for Solution to the Niger Delta Conflict: Some Proposals from a Conflict Transformation Approach." Memorandum submitted to the Technical Committee on the Niger Delta by the Federal Government of Nigeria, September 2008.

Owei, Opunebo, Samson George and Samuel Wachukwu. "Draft Report of the Faith Baptist Church Port Harcourt History Committee," March 2003.

Oyibu, Abigail. "Report on the Mission Trip to Opokuma, Bayelsa State by NDDC/AM Project in Collaboration with The Hilltop International Christian Centre, Port Harcourt," 6th-11th August, 2006.

Oyadongha, Samuel. "Five Feared Dead in Bayelsa Communal Clash." *Vanguard*, Lagos, http://www.vanguardngr.com/articles/2002/niger_delta/nd126042007.html (accessed April 25, 2007).

Pyagbara, Legborsi Saro. Minority Rights Group International, London. "The Ogoni of Nigeria: Oil and Exploitation," htpp://www.minorityrights.org (accessed February 9, 2006.

Ross, Michael L., UCLA Department of Political Science, Los Angeles. "Nigeria's Oil Sector and the Poor." Paper prepared for the UK Department for International Development "Nigeria: Drivers of Change" Program, May 23, 2003.

Sagay, Itse. "Nigeria: Federalism, the Constitution and Resource Control." Paper presented at the Fourth Sensitisation Programme Organised by the Ibori Vanguard at the Lagoon Restaurant, Lagos, n.d.

Subair, Gbola and Kayode Akinmade. "NNPC can't Account for ₦555bn - Revenue Commission." *The Tribune*, Lagos, August 13, 2007, http://www.tribune.com.ng/13082007/news/news1.html (accessed August 13, 2007).

Technical Committee on the Niger Delta. *Report of the Technical Committee on the Niger Delta*, Volume 1, November 2008.

Turaki, Yusufu. "Africa Traditional Religious System as Basis of Understanding Christian Spiritual Warfare," http://www.lausanne.org/Brix?pageID=13876 (accessed April 17, 2007).

Udeh, John. "Petroleum Revenue Management: The Nigerian Perspective." Paper Presented at a Workshop on Petroleum Revenue Management by Oil, Gas, Mining and Chemicals Department of WBG and ESMAP in Washington, D.C., on October 23-24, 2002.

Utomi, Pat. "The Curse of Oil." Paper Delivered at Heinrich Boll Foundation Oil-Conference, May 2004.

Van Eck, Ernest, Department of New Testament Studies, University of Pretoria. "The Word is Life: African Theology as Biblical and Contextual Theology." Lecture given at the opening of the academic year of the African Institute for Missiology (AIM) on 27 January 2005 at Klipdrif, Hammanskraal, South Africa.

Wokoma, Iyenemi Norman. "Assessing Accomplishments of Women's Nonviolent Direct Action in the Niger Delta," n.d.

Yar'Adua, Umaru Musa. "Amnesty Proclamation." Text of Amnesty Proclamation for Niger Delta Militants, June 25, 2009.

## Select Unpublished Theses and Dissertations

Brooks, L.J. "'Shell is the only Government We Know': Multi-sector Partnerships and Equitable Governance in the Niger Delta." PhD Thesis submitted to the London School of Economics and Political Science, 2004.

Ebeku, Kaniye Samuel Adheledhi. "Legal Aspects of Environmental Issues and Equity Considerations in the Exploitation of Oil in Nigeria's Niger Delta." PhD Thesis submitted to the Kent Law School (Faculty of Social Sciences), University of Kent at Cambridge, UK, September 2002.

Keeter, Gregory Todd. "Can Religion Help? Using John Howard Yoder and Mohandas Gandhi to Conceptualize New Approaches to Intractable Social and Political Problems Such as Violence and War." A Thesis Submitted in Partial Fulfillment of Requirements for the Degree of Master of Arts in the College of Arts and Sciences, Georgia State University, May 2006.

Muhammed-Jallo, Saidu. "The Development of the Petroleum Industry and Petroleum Law in Nigeria." Thesis submitted to the Centre for Petroleum and Mineral Law Studies, University of Dundee for the Award of LL.M Degree in Petroleum Law, April 1989.

Njeze, Cally C.O. "Oil Concessions and Land Acquisition in Nigeria." PhD Thesis submitted to the University of Cambridge, UK, 1978.

Onosode, Inerhunwa Christopher. "An Analysis of the Effects of the Oil Industry on Economic Development in Nigeria." PhD Thesis submitted to the University of Kent at Canterbury (Department of Economics), UK 1998.

Steyn, Maria Sophia. "Oil Politics in Ecuador and Nigeria: A Perspective from Environmental History on the Struggles between Ethnic Minority Groups, Multinational Oil Companies and National Governments." PhD Thesis submitted to the Faculty of Humanities (Department of History) University of the Free State, Bloemfontein, South Africa, 2003.

Turner, T.E. "Government and Oil in Nigeria: A Study of the Making and Implementation of Petroleum Policy." PhD Thesis submitted to the London School of Economics and Political Science, 1977.

## Select Nigerian Laws

Federal Republic of Nigeria, *Constitution of the Federal Republic of Nigeria, 1999.*

Federal Republic of Nigeria, *Environmental Impact Assessment Decree No. 86 of 1992.*

Federal Republic of Nigeria, *Land Use Act 1978.*

Federal Republic of Nigeria, *Oil Pipelines Act 1956.*

Federal Republic of Nigeria, *Minerals and Mining Decree No. 34 of 1999.*

Federal Republic of Nigeria, *Niger-Delta Development Commission (Establishment etc) Act No. 6 of 2000.*

Federal Republic of Nigeria, *Nigerian National Petroleum Corporation Act of 1977.*

## Select Newspapers

*BusinessDay,* Lagos

*The Economist,* London

*The Guardian*, Lagos

*The Sun,* Lagos

*The Tide*, Port Harcourt

*The Tribune,* Lagos

*ThisDay,* Lagos

*Vanguard*, Lagos